Lecture Notes in Artificial Intelligence 11326

Subseries of Lecture Notes in Computer Science

More information about this series at http://www.springer.com/series/1244

Fernando Koch · Andrew Koster
Isabelle Bichindaritz · Pau Herrero et al. (Eds.)

Artificial Intelligence in Health

First International Workshop, AIH 2018
Stockholm, Sweden, July 13–14, 2018
Revised Selected Papers

 Springer

Editors
Fernando Koch (iD)
University of Melbourne
Melbourne, VIC, Australia

Isabelle Bichindaritz (iD)
State University of New York
Oswego, NY, USA

Andrew Koster (iD)
IIIA-CSIC and Autonomous University
of Barcelona
Barcelona, Spain

Pau Herrero (iD)
Imperial College London
London, UK

Additional Workshop Editors *see next page*

ISSN 0302-9743   ISSN 1611-3349   (electronic)
Lecture Notes in Artificial Intelligence
ISBN 978-3-030-12737-4   ISBN 978-3-030-12738-1   (eBook)
https://doi.org/10.1007/978-3-030-12738-1

Library of Congress Control Number: 2019930650

LNCS Sublibrary: SL7 – Artificial Intelligence

This Springer imprint is published by the registered company Springer Nature Switzerland AG
The registered company address is: Gewerbestrasse 11, 6330 Cham, Switzerland

Additional Workshop Editors

David Riaño
Universitat Rovira i Virgili
Tarragona, Spain

Sara Montagna
Università di Bologna
Bologna, Italy

Michael Schumacher
University of Applied Sciences
and Arts Western Switzerland
(HES-SO)
Sierre, Switzerland

Annette ten Teije
Vrije Universiteit Amsterdam
Amsterdam, The Netherlands

Christian Guttmann
Nordic AI Institute, Karolinska Institute,
TIETO
Stockholm, Sweden

Manfred Reichert
Ulm University
Ulm, Germany

Richard Lenz
University of Erlangen
Erlangen, Germany

Beatriz López
University of Girona
Girona, Italy

Cindy Marling
Ohio University
Athens, OH, USA

Clare Martin
Oxford Brookes University
Oxford, UK

Stefania Montani
Università del Piemonte Orientale
Alessandria, Italy

Nirmalie Wiratunga
The Robert Gordon University
Aberdeen, UK

Preface

Artificial Intelligence is a key tool to evolve health systems by providing tools that aid medical professionals analyse available information in new ways, support patients with smart applications, and open new frontiers in diagnostics through the use of machine learning with unprecedented amounts of data. These are just some of the topics that were discussed during the Joint Workshop on Artificial Intelligence in Health (AIH 2018)[1], which consolidated the workshops CARE, KRH4C and AI4HC into a single event. The workshop took place in Stockholm, Sweden, July 13–14, 2018.

The objective of the workshop was to provide a discussion forum for the most recent and innovative work on the study and application of AI technologies in compelling health-care scenarios. The workshops covered a wide spectrum of applications, from those aimed at easing and supporting health-care professional work to those devoted to improving patient lives. The discussions revolved around AI technologies with medical applications focusing on three tracks: Agents in Health Care; Data Science and Decision Systems in Medicine; and Knowledge Management in Health Care. The workshop promoted an international discussion forum with Program Committee members from many countries in Asia (Israel), Europe (Germany, France, UK, Russia, The Netherlands, Switzerland, Spain, Italy, Norway, Sweden, Poland, Czech Republic, Greece and Portugal), Oceania (Australia and New Zealand), and the Americas (USA, Canada, Brazil and Colombia). A common theme through the tracks was the ethical use of AI in health care, around which there was a panel discussion with Dr. Christian Guttmann, Dr. Maite Lopez, and Dr. Anthony Chang.

The workshop received 42 submissions from which we selected 26 for presentation, and 18 extended versions were selected for the proceedings. All submissions were reviewed by at least three different reviewers. Papers being published in this volume

---

[1] AIH 2018: http://sots.brookes.ac.uk/~p0072382/ai4h2018/

highlight the innovation and contribution to the field, providing a review of the knowledge domain, challenges, opportunities, and contributions to real-world problems.

December 2018

Fernando Koch
Andrew Koster
Isabelle Bichindaritz
Pau Herrero
David Riaño
Sara Montagna
Michael Schumacher
Annette ten Teije
Christian Guttmann
Manfred Reichert
Richard Lenz
Beatriz López
Cindy Marling
Clare Martin
Stefania Montani
Nirmalie Wiratunga

Organization

Organizing Committee

Isabelle Bichindaritz	State University of New York at Oswego, USA
Christian Guttmann	Nordic AI Institute, Karolinska Institute, TIETO, University of New South Wales, Sweden and Australia
Pau Herrero	Imperial College London, UK
Fernando Koch	IBM Global Services, USA, and The University of Melbourne, Australia
Andrew Koster	IIIA-CSIC and Autonomous University of Barcelona, Spain
Richard Lenz	University of Erlangen, Germany
Beatriz López	University of Girona, Spain
Cindy Marling	Ohio University, USA
Clare Martin	Oxford Brookes University, University of Buckingham, Tesssella Support Services, University of Oxford, UK
Sara Montagna	Università di Bologna, Italy
Stefania Montani	Università del Piemonte Orientale, Italy
Manfred Reichert	Ulm University, Germany
David Riaño	Universitat Rovira i Virgili, Universitat Politècnica de Catalunya, Spain
Michael Schumacher	University of Applied Sciences and Arts Western Switzerland (HES-SO), Switzerland
Annette ten Teije	Vrije Universiteit Amsterdam, The Netherlands
Nirmalie Wiratunga	The Robert Gordon University, UK

Technical Program Committee

Agnar Aamodt	Norwegian University of Science and Technology, Norway
Syed Sibte Raza Abidi	Dalhousie University, Canada
Klaus-Dieter Althoff	DFKI/University of Hildesheim, Germany
Luca Anselma	Università di Torino, Italy
Josep Lluis Arcos	IIIA - CSIC, Artificial Intelligence Research Institute
Joseph Barjis	San Jose State University, USA
Isabelle Bichindaritz	State University of New York at Oswego, USA
Davide Calvaresi	University of Applied Sciences Western Switzerland, Switzerland
Michel Dojat	INSERM, France
Aldo Franco Dragoni	Università Politecnica delle Marche, Italy
Néstor Darío Duque Méndez	Universidad Nacional de Colombia, Colombia

Christian Guttmann	Nordic AI Institute, Karolinska Institute, TIETO, University of New South Wales, Sweden and Australia
Pau Herrero	Imperial College London, UK
Alec Holt	University of Otago, New Zealand
David Isern	Universitat Rovira i Virgili, ITAKA Research Group, Spain
Stefan Jablonski	University of Bayreuth, Germany
Fernando Koch	IBM Global Services, USA, and The University of Melbourne, Australia
Andrew Koster	IIIA-CSIC and Autonomous University of Barcelona, Spain
Vassilis Koutkias	Institute of Applied Biosciences, Centre for Research and Technology Hellas, Greece
Richard Lenz	University of Erlangen, Germany
Lenka Lhotska	Czech Technical University in Prague, Czech Republic
Jean Lieber	LORIA - Inria Lorraine, France
Beatriz López	University of Girona, Spain
Mar Marcos	Universitat Jaume I, Spain
Cindy Marling	Ohio University, USA
Clare Martin	Oxford Brookes University, UK
Sara Montagna	Università di Bologna, Italy
Antonio Moreno	URV
Juan Carlos Nieves	Umeå University, Sweden
Øystein Nytrø	Norwegian University of Science and Technology, Norway
Stefan Pantazi	Conestoga College, Canada
Hugo Paredes	INESC TEC and UTAD
Mor Peleg	University of Haifa, Israel
Petra Perner	Institute of Computer Vision and Applied Computer Sciences
Luigi Portinale	Università del Piemonte, Italy
Tiago Primo	Federal University of Pelotas, Brazil
Manfred Reichert	University of Ulm, Germany
David Riaño	Universitat Rovira i Virgili, Spain
Sadiq Sani	The Robert Gordon University, UK
Rainer Schmidt	University of Rostock, Germany
Michael Schumacher	University of Applied Sciences and Arts Western Switzerland (HES-SO), Switzerland
Brigitte Séroussi	Assistance Publique - Hôpitaux de Paris, France
Jaime Sichman	University of São Paulo, Brazil
Maria Taboada	University of Santiago de Compostela, Spain
Annette ten Teije	Vrije Universiteit Amsterdam, The Netherlands
Paolo Terenziani	Università del Piemonte Orientale, Italy
Ingo J. Timm	University of Trier, Germany
Frank Van Harmelen	Vrije Universiteit Amsterdam, The Netherlands
Eloisa Vargiu	Eurecat Technology Center - eHealth Unit

Olga Vorobieva   Instiute of Evolutionary Physiology and Biochemistry
Dongwen Wang   Arizona State University, USA
Szymon Wilk   Poznan University of Technology, Poland
Nirmalie Wiratunga   The Robert Gordon University, UK

Conference Committee

Agnar Aamodt   Norwegian University of Science and Technology,
                          Norway
Syed Sibte Raza Abidi   Dalhousie University, Canada
Klaus-Dieter Althoff   DFKI/University of Hildesheim, Germany
Luca Anselma   Università di Torino, Italy
Josep Lluis Arcos   IIIA - CSIC, Artificial Intelligence Research Institute,
                          Spain
Joseph Barjis   San Jose State University, USA
Isabelle Bichindaritz   State University of New York at Oswego, USA
Davide Calvaresi   University of Applied Sciences Western Switzerland,
                          Switzerland
Michel Dojat   INSERM, France
Aldo Franco Dragoni   Università Politecnica delle Marche, Italy
Néstor Darío Duque   Universidad Nacional de Colombia, Colombia
  Méndez
Christian Guttmann   Nordic AI Institute, Karolinska Institute, TIETO,
                          University of New South Wales, Sweden and Australia
Pau Herrero   Imperial College London, UK
Alec Holt   University of Otago, New Zealand
David Isern   Universitat Rovira i Virgili, ITAKA Research Group,
                          Spain
Stefan Jablonski   University of Bayreuth, Germany
Fernando Koch   IBM Global Services, USA, and The University
                          of Melbourne, Australia
Andrew Koster   IIIA-CSIC and Autonomous University of Barcelona,
                          Spain
Vassilis Koutkias   Institute of Applied Biosciences, Centre for Research
                          and Technology Hellas, Greece
Richard Lenz   University of Erlangen, Germany
Lenka Lhotska   Czech Technical University in Prague, Czech Republic
Jean Lieber   LORIA Lorraine, France
Beatriz López   University of Girona, Spain
Mar Marcos   Universitat Jaume I, Spain
Cindy Marling   Ohio University, USA
Clare Martin   Oxford Brookes University, UK
Sara Montagna   Università di Bologna, Italy
Stefania Montani   Università del Piemonte Orientale, Italy
Antonio Moreno   URV
Juan Carlos Nieves   Umeå University, Sweden

Acknowledgments

We would like to thank all the volunteers who made the workshops possible by helping to organise and peer review the submissions, and EasyChair for the conference and proceedings management system. We appreciate the help and dedication of the members of the AIH research communities, who continuously participate in our activities as Technical Program Committee, submitting contributions, and helping us put the pieces together to promote this publication.

We are also grateful to Springer for the continuous support and providing the venue for publish the proceeding after our workshops. This contributing in invaluable to further promoting the research around agents in health care, data science and decision systems in medicine, and knowledge management in health care.

During the implementation of this project, Dr. Koch was working with the School of Computing and Information Systems, The University of Melbourne, sponsored by a Global Research Outreach grant from Samsung Advanced Institute of Technology, collaboration Project IO170924-04695-01. Dr. Koch is also supported by the CNPq Productivity in Technology and Innovation Award, grant CNPq 307275/2015-9. Currently, Dr. Koch works for IBM Global Services, Armonk, NY, USA.

Dr. Koster is a Marie Skłodowska-Curie fellow at the Artificial Intelligence Research Institute (IIIA-CSIC) and the Autonomous University of Barcelona, funded by the European Commission's Horizon 2020 research and innovation programme under grant agreement 665919. Dr. Koster also thanks the Generalitat de Catalunya (Grant: 2014 SGR 118).

Dr. David Riaño is associate professor of the Department of Computer Science and Mathematics at Universitat Rovira i Virgili (URV) in Tarragona, Spain. He is member of the board of the Society for Artificial Intelligence in Medicine (AIME) and head of the Research Group on Artificial Intelligence for Health Care at URV. He is sponsored by the RETOS P-BreasTreat project (DPI2016-77415-R) of the Spanish Ministerio de Economia y Competitividad.

Dr. Montagna is a Research Fellow at the Department of Computer Science and Engineering (DISI) of the University Bologna since October 2012. She thanks the DISI Department for the budget (BIR) devoted to co-finance research projects.

Dr. Guttmann has been an adjunct associate professor at the University of New South Wales, a researcher at the Karolinska Institute, the executive director of the Nordic AI Institute and Vice President and the Chief AI Scientist at Tieto.

Prof. Reichert is full professor and director of the Institute of Databases and Information Systems at Ulm University, Germany. He is member of the Management Board of the European School for Interdisciplinary Tinnitus Research funded by the European Union's Horizon 2020 research and innovation programme under the Marie Skłodowska-Curie grant agreement number 722046. Furthermore, he participates in the European Joint Action CHRODIS+, which aims at implementing sustainable practices for chronic diseases.

Dr. Bichindaritz is an associate professor in computer science and Director of biomedical informatics at State University of New York in Oswego. Her research focuses on intelligent learning systems and biomedical data science for decision analytics and data analytics in health care and biomedical research, as well as more broadly artificial intelligence in medicine. She has organized or co-organized many international workshops, including Data Mining in Life Sciences, Case-based Reasoning in the Health Sciences, and Synergies between Case-based Reasoning and Machine Learning. She thanks the State University of New York for funding her work in biomedical informatics through a number of grants.

Dr. Pau Herrero is a Research Fellow at the Department of Electrical and Electronic Engineering, Imperial College London (UK), funded by the European Commission (H2020 programme) under grant agreement 689810-PEPPER.

Prof. Richard Lenz is a full professor for data management at the University of Erlangen-Nuremberg, Germany. He is doing research in evolutionary information systems, information systems in health care, data quality and data integration, and process management.

Dr. Beatriz López is associate professor at the University of Girona, Spain. Her research is partially funded by the European Commission (H2020 programme) under grant agreement 689810-PEPPER. Dr. Lopez thanks the Generalitat de Catalunya (Recognition award: 2017 SGR 1551), and the University of Girona (Grant MPCUdG2016 and GdRCompetudG2017).

Dr. Cindy Marling is an associate professor in the School of Electrical Engineering and Computer Science, Adjunct Associate Professor in the Department of Specialty Medicine, and Director of the SmartHealth Lab at Ohio University. Her research is supported by grant 1R21EB022356 from the National Institutes of Health (NIH). Dr. Clare Martin is a Principal Lecturer at Oxford Brookes University, UK. Her research is funded by the European Commission (H2020 programme) under grant agreement 689810-PEPPER.

Dr. Stefania Montani is a full professor in Computer Science at DISIT, Computer Science Institute, University of Piemonte Orientale in Alessandria, Italy. Her research interests are related to decision support and process management, with a focus on medical applications. She thanks the University of Piemonte Orientale for funding her recent work in medical informatics.

Dr. Nirmalie Wiratunga is a Professor at the School of Computer Science and Digital Media, Robert Gordon University, Aberdeen, UK. Her research is funded by the European Union Horizon 2020 research and innovation programme under grant agreement 689043 and also InnovateUK grant (1024226), EPSRC Network+ (GetAMoveOn).

Contents

Data Science and Decision Systems in Medicine

Agents in Health Care and Knowledge Management in Health Care

Part I: Agents in Health Care and Knowledge Management in Health Care

The first part of this volume includes extended and revised versions of a set of selected work from the track "Agents in Health Care and Knowledge Management in Health Care".

In the first chapter, "MeSHx-Notes: Web-System for Clinical Notes", Nunes *et al.* outline a novel application for automatically highlighting and describing medical terms in clinical notes, in order to make it easier and faster to understand these texts. The presentation includes an early-stage demonstration of the proof-of-concept implementation along with an evaluation of the proposal's applicability.

The second chapter, "Multiagent Systems to Support Planning and Scheduling in Home Health Care Management: A Literature Review", Becker *et al.* provides an overview of operational management processes applied for planning and scheduling. The work focuses on the home health care systems application domain. The paper presents a systematic literature review on relevant approaches, along with a comparative analysis and identified deficiencies. The work concludes on the lack of an all-encompassing approach and infers the opportunity for contributing with new methods of dynamic distributed scheduling for the control of operational processes to increases efficiency and optimise resource utilisation in these environments.

In the third chapter, "Ethical Surveillance: Applying Deep Learning and Contextual Awareness for the Benefit of Persons Living with Dementia", Williams, Ware and Müller propose a privacy-aware system for tracking when persons with dementia exhibit risky behaviour. The innovation revolves around a system that prevents third-parties spying on the subject. The method employs machine learning techniques to detect if the subject is at risk and if there is sufficient danger. The application is designed to notify caretakers of the subject's location.

In the fourth chapter, "Active Learning for Conversational Interfaces in Health-Care Applications", Härmä, Polyakov and Chernyak present a novel method for choosing which samples to manually label in a large and problematic data set of conversation. The work compares the proposed method *query by embedded committee* with the state of the art and demonstrate positive results in providing similarly accurate classifiers with fewer labeled samples.

In the fifth chapter, "Analysis of Topic Propagation in Therapy Sessions Using Partially Labeled Latent Dirichlet Allocation", Chaoua *et al.* present an innovation method combining topic modeling and transitions matrices to detect and track topics in real-life psychotherapeutic conversation. The method is based on Partially Labelled Dirichlet Allocation and allows to identify the semantic themes of the current therapeutic conversation and predict topics for each talk-turn between the patient and the counsellor. In addition, the method also proposes a solution to understand and explore the dynamics of the conversation giving insights and tips on logic and strategy to adopt.

Next, Raff, Lantzy and Maier present the discussion "Dr. AI, Where did you get your degree?" on regulatory issues surrounding AI in medicine. They propose a novel approach by treating sufficiently advanced diagnostic tools more like medical professionals and less like medical tools. The paper reviews several issues related to this approach, introduces a regulatory framework, and foments the discussion of how medical AI in medicine may be regulated.

In the seventh chapter, "Principles for agent-based assistive technology design", Guerrero *et al.* describe a process for deciding when the technology should take action, and when a patient or a caregiver able to perform the task without further help. They use formal argumentation to create a framework for reasoning about situations and use the concept of zones of proximal development to make decisions. The paper presents a proof-of-concept implementation applying augmented reality and discusses utilisation issues.

The eighth chapter, "Microsoft Hololens, a mHealth Solution for Medication Adherence", Ingeson, Blusi and Nieves introduce a smart mHealth application based on the augmented reality to support patients with management of their medication. The solution allows evaluating patients' performance over time and adapt in order to improve interventions. The work presents a proof-of-concept implementation with the solution embodied as a smart augmented reality application on top of Microsoft HoloLens, testing in a controlled environment. The results were very positive demonstrating the potential to utilise the proposed technology in real-world settings.

Completing this first part, chapter "A Knowledge-Based Simulation Framework for Decision Support in Brazilian National Cancer Institute", by Gonçalves *et al.* present a knowledge-based simulation framework developed at the Brazilian National Cancer Institute (INCA) to reduce patients' waiting time to start cancer treatment. The system evaluates *what-if* scenarios to identify potential negative cases. The solution is being tested in a controlled environment leading to optimisation of waiting time for cancer treatment, what impacts on patients' quality of life.

MeSHx-Notes: Web-System for Clinical Notes

Rafael O. Nunes, João E. Soares, Henrique D. P. dos Santos[(⊠)][iD],
and Renata Vieira[iD]

School of Technology at Pontifical Catholic University of Rio Grande do Sul,
Porto Alegre, Brazil
{rafael.oleques,joao.etchichury,henrique.santos.003}@acad.pucrs.br,
renata.vieira@pucrs.br

Abstract. We present MeSHx-Notes, MeSH eXtended for clinical notes,
a multi-language web system based on the Django framework to present
selected terms in clinical notes. MeSHx-Notes extends Medical Subject
Headings (MeSH) terms with Word Embeddings with similar words.
Since MeSH is available in 15 languages, MeSHx-Notes is easily extend-
able by replacing the MeSH thesaurus with the target language (plus the
generation of the corresponding WE for the new language). Our version
deals with Portuguese and English.

Keywords: Multi-language · Web system · Clinical notes ·
Information extraction · Word Embeddings · MeSH

1 Introduction

Electronic Health Records (EHR) play an important role in hospital environments, bringing many benefits in terms of patient safety, satisfaction, and effec-
tiveness/efficiency of care [1]. Records of health care practices in hospitals generate a rich and large amount of patient information and an intrinsic relation
between symptoms, diseases, drug interactions, and diagnoses that may be used
for many purposes [2,7,8]. Clinical notes, such as discharge summaries, have
a semi- or unstructured format. These documents contain information about
diseases, treatments, drugs, etc. Extracting meaningful information from them
becomes challenging due to their narrative format [5].

This work aims to help healthcare professionals concerning the understanding of what is informed in clinical notes. This is possible through the use of
Natural Language Processing (NLP), combined with the MeSH dictionary[1]. We
developed a web application that exhibits the meaning and the related words
for terms of a set of categories used in clinical notes, thus enhancing the understanding of what is reported.

---

[1] https://www.ncbi.nlm.nih.gov/mesh.

© Springer Nature Switzerland AG 2019
F. Koch et al. (Eds.): AIH 2018, LNAI 11326, pp. 5–12, 2019.
https://doi.org/10.1007/978-3-030-12738-1_1

In this context, we present an easy-to-use system that provides users with extra knowledge of the information given in clinical notes, which can be used by anyone with access to the internet.

The rest of this paper is organized as follows: Sect. 2 presents previous works on information extraction through clinical notes. In Sect. 3 we explain the concepts related to the term expansion. Section 4 describes the concepts used in the MeSHx-Notes system, followed by the results in Sect. 5. Finally, in Sect. 6 we summarize our contributions and present further research directions.

2   Related Work

One problem in clinical notes is that registers are not always is accordance with the standard language, therefore the identification of the right dictionary entry is challenging [12]. Clinical notes usually contain abbreviations, misspelled words, and word concatenations. To overcome such problems, we propose the use of Word Embedding models (generated on the basis of clinical notes) to spot terms that are similar to the dictionary entries.

The use of pre-established ontologies for the classification of medical documents has also become a trend, since such structures already bring a semantic knowledge of the data and help in the organization of texts [10]. The US National Library of Medicine has developed an ontology for medical systems to communicate, called the Unified Medical Language System (UMLS). The same project includes the medical subject ontology, known as MeSH, which relates the medical vocabularies of diseases, symptoms, organs, etc.

While other systems, such as cTAKES [9], rely on several UMLS sources for English to provide several information from clinical notes, we focus on developing a user-friendly and easy-to-handle web interface, portable for languages other than English, using a language-specific MeSH thesaurus.

Several efforts have been reported in the area of clinical text mining to bridge the gap between unstructured clinical notes and structured data representation, including tools such as MetaMap and KnowledgeMap, which have been developed to automatically annotate medical concepts in free text, along with systems to identify the patient's disease status, medication information, etc. [3].

3   MeSH Dictionary Expansion

3.1   Medical Subject Headings (MeSH)

The MeSH dictionary (Medical Subject Headings) (see footnote 1) is the National Library of Medicine controlled vocabulary thesaurus used to index articles for PubMed.

Started in 2013, MeSH has 54,935 entries where each entry has a unique tree number and consists of 26,851 main headings and 213,000 entry terms that increase the power of classification of medical documents. MeSH is available in 15 languages: English, Croatian, Czech, Dutch, Finnish, French, German, Italian, Japanese, Norwegian, Polish, Portuguese, Russian, Spanish, and Swedish.

In MeSH, each heading has information about a term - e.g., Unique ID, Scope Note, qualifiers, and Entry Terms. The unique ID refers to the working term, therefore, homonyms "perna" (leg) and "perna" (organism) have distinct IDs. The Scope Note refers to the term's meaning. Qualifiers divide terms into categories. Entry Terms are synonymous or alternative ways to write a term, for instance "ache" is an Entry Term of "pain".

MeSH has 81 qualifiers. In our study we selected five categories: pharmacology, anatomy, methods, diagnosis, and others. These are the most frequent qualifiers, according to our analysis. The category 'others' includes the least frequent 77 categories and terms that do not have a qualifier. We use the qualifiers as a way to classify not only terms previously found in MeSH but also the new terms with which our dictionary was enriched, as explained below.

3.2   Electronic Health Records

The Portuguese dataset was obtained from Hospital Nossa Senhora da Conceição (HNSC). We used a large cohort extracted from the administrative hospitalization database from this Hospital. HNSC is part of the Brazilian public healthcare system and provides tertiary care. The data comprises 1.5 million clinical notes from 48.9 thousand hospitalization records annotated with the Charlson comorbidity index between January 2012 and December 2017.

Ethical approval to use the hospital dataset in this research was granted by the Research Ethics Committee of Conceição Hospital Group under the number 71571717.7.0000.5530.

The English dataset was obtained from i2b2 Challenge [11] from 2008 to 2012. It is a set of nine datasets from several shared tasks promoted by Informatics for Integrating Biology and the Bedside (i2b2). In the 2012 i2b2 Challenge, 310 discharge summaries were annotated for temporal information. The challenge focused specifically on the identification of clinically relevant events in patient records and on the relative ordering of the events with respect to each other and with respect to time expressions included in the records.

3.3   Word Embeddings

Word vectors are a way of mapping words in a numerical space. A latent syntactic/semantic vector for each word is induced from a large unlabeled corpus. The Portuguese and English model for the word embeddings was trained with Word2Vec [4]. For the Portuguese version, we used 21 million sentences from HNSC's medical records, trained with 50 dimensions per word and a minimum word count of 100 [6]. This training resulted in 73 thousand word vectors. For the English version, we used 171 thousand sentences from the i2b2 challenge dataset, trained with 50 dimensions and a minimum word count of 10, resulting in 17 thousand word vectors.

The original dictionary was expanded using Word Embeddings. The expansion process was made by analyzing the similarity of the MeSH's Entry Terms of each input with those from the Word Embeddings. The terms which were considered similar were linked to the specific Unique ID of the enriched dictionary and added to a reverse index.

Table 1. Enrichment of MeSH terms

Heading	Original terms	New similar terms
Abdomem	abdomem, belly	abd, abdome...
Celecoxib	celecoxib, celebrex	norvasc, losartan...
Abscess	abcesso, absceso	abscess, abscesses...

Table 1 shows some examples of heading terms in the MeSh dictionary, their alternative terms, and the corresponding new identified terms. For example, the heading "Abscesso" had "abcesso" and "absceso" as alternative MeSH terms, and "abscess" and "abscesses" were added as new terms found through the WE model. Originally the dictionary had 80,973 terms; with the expansion there was an enrichment of 40,588 new terms. The enrichment brings new terms due to abbreviations, orthography errors, and word concatenations. Table 2 shows examples of such cases.

Table 2. Enriched dictionary terms

MeSH terms	Expanded terms
Tomography	tomo, tc, tomographyexanms
General surgery	srg, surrgery, sugery
Enoxaparin	enoxa, enoxeparin
Fever	chills, hyperthermia
Behavior	behav, bhv

4 MeSHx-Notes: System Description

The system consists of a web application that receives clinical notes, identifies the main terms, and then returns their definition, similar words and a link to the MeSH dictionary. Its development is based on Python, Django, Pandas, Bootstrap, JQuery, Word Embeddings, XPath, and the MeSH thesaurus.

In the web page, buttons are provided to navigate between clinical notes and to change the language. Besides, the clinical note description is given, with data

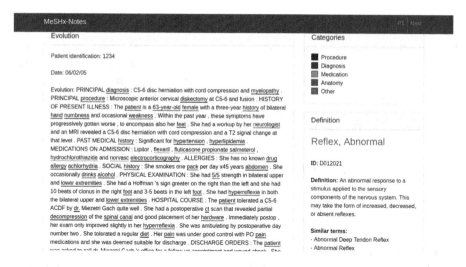

Fig. 1. Example of clinical notes with highlighted terms, color legend for each category, and MeSH descriptor.

about the patient record and its modification date with a concomitant section of legends that are related to the classification of the terms. Nonetheless, identified words are underlined according to their classification, so that, when clicked, they show their technical name, ID, description, terms with similar meanings, and a link to the MeSH description website. Some special features were implemented, such as highlighting of terms of a specific category.

Having in mind the processing time for information to be presented to the user, we search for terms using a reverse index, previously generated with the terms and their IDs. MeSHx-Notes was built for Portuguese and English.

This system shows the definition of medical terms, helping in their understanding. It can be applied in various situations: some applications improve learning of nursing, medical or other health-related students, in addition to aiding multidisciplinary research groups in which not all members have technical medical knowledge. Besides, the system works not only with clinical notes, but also with any texts related to health, for example, journalistic and academic texts.

4.1 Back-End

First, the extended MeSH dictionary is generated, using previously saved data in an XML file, containing ID, name, scope, terms, and qualifier. The dictionary is enriched to provide a greater range of terms, which are stored in the terms field. We consider higher similarity degrees to identify those words.

After that step, we read the clinical notes, using Pandas in the web application, using Django as the development framework. Each word found in the dictionary is captured, and the lists of original and new similar words are stored.

4.2   Front-End

When a clinical note is shown to the user, terms from the (enriched) dictionary are highlighted. These words are shown in different colors, according to the following classes: medication, diagnosis, procedure or anatomy. It is possible to select specific classes, providing better information visualization (e.g., only medication to look for what the patient is using in a treatment). This development used JQuery and Bootstrap.

5   Term Expansion Evaluation

For the expansion of alternative terms, there was a manual evaluation of the enriched terms from 42 clinical notes, whose expanded terms were annotated as "correct" or not. Based on this evaluation, a gold standard was generated with all the appended terms and their Unique IDs, to which each term should be properly related. As a result, we had 651 examples. Based on that, we tested several thresholds, to estimate the best threshold for each category.

Through the gold standard and different thresholds, we obtained values with the lowest failure rates using an algorithm. This algorithm analyzed the accuracy of each threshold, as shown in Table 3. The analysis started with a threshold of 0.80 to 0.99–1.00 returns the term itself. Thus, we accomplished a 58% accuracy rate assessing 691 terms contained in our gold standard. The obtained results are based on tests with real clinical notes.

This accuracy value of 58% is yet to be improved. At this stage, the user still has to judge for themselves the alternatives presented by the system. However, given the difficulty of the task, we consider that this initial result is promising and there are ways in which it could be further improved.

Table 3. Best thresholds per qualifier

Qualifier	Thresholds	Correct	Total	Accuracy
Methods	0.89	80	116	71%
Diagnosis	0.93	81	164	49%
Pharmacology	0.96	137	263	65%
Anatomy	0.95	44	83	54%
Others	0.94	97	188	51%

True positives (correctly enriched terms) are presented in Table 4. There are terms with lexical similarity (e.g., rehab and rehabilitation), but also terms that are semantically similar but lexically distant—e.g., amlodipine and norvasc.

Table 4. True positive terms

MeSH term	True positive
Arteries	Vessels
Angioplasty	Stenting
Rehabilitation	Rehab
Amlodipine	Norvasc

Table 5 presents examples of false positives, terms that were incorrectly identified as similar through Word Embeddings. These terms, on the established
threshold, had a degree of similarity. New NLP techniques, based on word-sense
disambiguation, are being studied to try to solve these problems.

Table 5. False positive terms

MeSH term	False positive
Thoracotomy	Parietal
Bicuspid	Ulcerative colitis
Ocular vision	Weakness

6   Conclusion and Further Work

MeSHx-Notes aims to provide, both for health professionals and for non-
specialists, a simple tool that enables a better understanding of the terms used
in clinical notes in a clear, concise, accessible way. The source code is available on the project's Github page[2]. A web demo is also available[3]. As further
work, we plan to use bigram and trigram embeddings to find similar multi-word
expressions.

Aiming to improve the system and the accuracy rate (58%), we will use
new disambiguation techniques and similarity analysis, besides the evaluation of
enriched terms made by nurses. MeSH ambiguity is a problem to be studied in
the continuation of this work. User pilot studies are an important phase to be
pursued to test whether the system enhances the readability of medical notes,
after we achieve better accuracy rates.

Another goal is to perform classification tasks in clinical notes written in Portuguese using MeSH codes. Then, we will validate the learned model in clinical
notes in English using MeSH terms for those codes. These experiments intend
to evaluate the cross-language ability of MeSH for classification tasks in different languages. Furthermore, our purpose is to use a new database with the
MeSH definitions in Portuguese [13]. This way, we will be able to better identify
ambiguous terms through their definition elements in clinical notes.

---

[2] https://github.com/nlp-pucrs/meshx-notes.
[3] http://grupopln.inf.pucrs.br/meshx.

Acknowledgments. This work was partially supported by CAPES (Coordenação de Aperfeiçoamento de Pessoal de Nível Superior) Foundation (Brazil), PUCRS (Pontifical Catholic University of Rio Grande do Sul), and UFRGS (Federal University of Rio Grande do Sul).

References

1. Buntin, M.B., Burke, M.F., Hoaglin, M.C., Blumenthal, D.: The benefits of health information technology: a review of the recent literature shows predominantly positive results. Health Aff. **30**(3), 464–471 (2011)
2. Jensen, P.B., Jensen, L.J., Brunak, S.: Mining electronic health records: towards better research applications and clinical care. Nat. Rev. Genet. **13**(6), 395 (2012)
3. Kovačević, A., Dehghan, A., Filannino, M., Keane, J.A., Nenadic, G.: Combining rules and machine learning for extraction of temporal expressions and events from clinical narratives. J. Am. Med. Inform. Assoc. **20**(5), 859–866 (2013)
4. Mikolov, T., Sutskever, I., Chen, K., Corrado, G.S., Dean, J.: Distributed representations of words and phrases and their compositionality. In: Advances in Neural Information Processing Systems, pp. 3111–3119 (2013)
5. Reátegui, R., Ratté, S.: Comparison of metamap and ctakes for entity extraction in clinical notes. BMC Med. Inform. Decis. Mak. **18**(3), 74 (2018)
6. dos Santos, H.D.P., Nunes, R.O., Soares, J.E., Vieira, R.: Meshx-notes: web system for clinical notes information extraction. In: AIH Joint Workshop on Artificial Intelligence for Health, p. 1. Stockholm, Sweden, July 2018
7. dos Santos, H.D.P., Ulbrich, A.H.D.P.S., Woloszyn, V., Vieira, R.: DDC-outlier: preventing medication errors using unsupervised learning. IEEE J. Biomed. Health Inform., 1 (2018). https://doi.org/10.1109/JBHI.2018.2828028
8. dos Santos, H.D.P., Ulbrich, A.H.D.P.S., Woloszyn, V., Vieira, R.: An initial investigation of Charlson comorbidity index regression based on clinical notes. In: 31st IEEE CBMS International Symposium on Computer-Based Medical Systems (CBMS), pp. 6–11. IEEE, Karlstad, June 2018. https://doi.org/10.1109/CBMS.2018.00009
9. Savova, G.K., et al.: Mayo clinical text analysis and knowledge extraction system (cTAKES): architecture, component evaluation and applications. J. Am. Med. Inform. Assoc. **17**(5), 507–513 (2010)
10. Trieschnigg, D., Pezik, P., Lee, V., De Jong, F., Kraaij, W., Rebholz-Schuhmann, D.: Mesh up: effective mesh text classification for improved document retrieval. Bioinformatics **25**(11), 1412–1418 (2009)
11. Uzuner, Ö., South, B.R., Shen, S., DuVall, S.L.: 2010 i2b2/VA challenge on concepts, assertions, and relations in clinical text. J. Am. Med. Inform. Assoc. **18**(5), 552–556 (2011)
12. Wang, Y., et al.: Clinical information extraction applications: a literature review. J. Biomed. Inform. **77**, 34 – 49 (2018). https://doi.org/10.1016/j.jbi.2017.11.011. http://www.sciencedirect.com/science/article/pii/S1532046417302563
13. Who, B.P.: Health sciences descriptors: DECS (2017). http://decs.bvsalud.org/I/homepagei.htm. Accessed 30 Sept 2018

Multiagent Systems to Support Planning and Scheduling in Home Health Care Management: A Literature Review

Colja A. Becker$^{(\boxtimes)}$(iD), Fabian Lorig(iD), and Ingo J. Timm(iD)

Business Informatics I, Trier University, Behringstrasse 21, 54296 Trier, Germany
{beckerc,lorigf,itimm}@uni-trier.de

Abstract. Ensuring sustainable care-giving systems with a focus on human needs and desires is a major challenge. An increasing demand in home health care as well as a limited number of professionals in the labor market have led to a call for efficiency. Thus, managing existing resources has gained relevance. The overall goal is high quality care services, while ensuring economic viability. At the same time, there is a need for modern customer-friendly solutions as well as the consideration of employees' preferences. To achieve this, adequate methods are needed that take current and future developments into account. Operational management processes in terms of planning and scheduling can be supported by multiagent systems as well as decision support systems using agent-based simulation. The aim of this work is to provide an overview of these solutions in the domain of home health care systems. To this end, we conducted a systematic literature review in which 11 relevant approaches were identified. In addition, these publications were analyzed to identify deficiencies and compared to each other. Because none of the approaches offers a sufficient solution, future work will focus on dynamic distributed scheduling for the control of operational processes which increases efficiency and improves the use of limited resources.

Keywords: Multiagent systems · Agent-based simulation ·
Logistics · Home health care · Scheduling · Software agents ·
Operations management · Multiagent planning ·
Distributed artificial intelligence

1 Introduction

Demographic change and urbanization have resulted in an increasing demand for care services. Decreasing birth rates and improved health care cause a higher ratio of elderly people, who potentially become care dependent. Due to globalization and an increasing willingness of younger people to relocate, relatives who could provide care might not be available. Furthermore, it is possible that the employment status of relatives does not allow them to provide intra-familial care services. Hence, it can be assumed that there is an increasing trend in demand

© Springer Nature Switzerland AG 2019
F. Koch et al. (Eds.): AIH 2018, LNAI 11326, pp. 13–28, 2019.
https://doi.org/10.1007/978-3-030-12738-1_2

for professional care services. In the near future, any decrease in the tendency of this development can not be expected, meaning that ensuring sustainable care-giving systems with focus on human needs and desires is a major challenge.

Instead of receiving care services in specialized care facilities, many care-receivers prefer to stay in their familiar environment. Such services are offered by home health care (HHC) service providers. The caregivers are equipped with cars and render the required care services in the respective patients' homes. By this means, care dependent persons receive the required assistance while maintaining their current way of living. To cope with an increasing demand in home health care, additional caregivers must be hired by service providers. However, in the labor market, qualified caregivers can be considered to be a limited rare resource. Following this, managing existing human resources in HHC gains in relevance to enable efficient service delivery. In this regard, the cost-benefit ratio of provided care services must be traded off against ethical aspects of care. The overall goal is to provide high-quality care services, while ensuring economic viability.

At the same time, modern customer-friendly solutions as well as the consideration of employees' preferences are required. Methods from *classic* HHC logistics focus only on scheduling and routing of caregivers. Yet, these methods do not seem sufficient with regard to *modern* HHC logistics, where in addition the individual desires of both caregivers and care recipients are considered as well as interaction between the participants. This includes the flexible adjustment of individual tasks or schedules for adaptively dealing with a dynamic environment. Moreover, taking real-world data into account might be necessary to achieve a proper planning result (e.g., traffic delay data). This also allows for dynamic changes of preferences: On the one hand, caregivers can for instance receive flexible schedules and are able to react on planning disturbances. On the other hand, care receivers are for example able to alter appointment time windows and demanded care services in the short term. Furthermore, management instructions should not only define or designate the tasks, but also define the scope of action based on individual qualifications, preferences, and other personal attributes of each employee. By this means, individuals are provided with both instructions on what tasks they have to accomplish and instructions regarding flexibility in their execution (e.g., sequence of accomplishment, type and manner of execution, as well as individual adaption of a task).

From an HHC provider's perspective, the management of this situation is challenging. Adequate methods are required that take current and future developments into account. To allow for corresponding management with focus on planning and scheduling, resulting requirements can be summarized as a need for flexibility in caregivers' operations, efficiency in the use of resources, and economic viability under present and future conditions. Considering these requirements, it is questionable whether and to what extent they are met by current approaches. In case no satisfying methods can be identified, as a first step, the question of shortcomings arises. To close this gap, the goal of this paper is to provide an overview of current approaches which support operational management in terms planning and scheduling in home health care systems. Moreover, if current approaches show deficiencies, these shortcomings must be analyzed and resulting challenges derived. To this end, a systematic literature review is conducted in order to gather relevant contributions.

The remainder of the paper is structured as follows: Sect. 2 presents background information. Section 3 introduces the methodology pursued in the systematic literature review as well as the applied search and evaluation criteria. Section 4 gives an overview of the surveyed approaches and Sect. 5 describes shortcomings of these approaches. Finally, the conclusion of this article and comments about further work are provided.

2   Background

To increase the efficiency of the operational management processes as well as the managed processes, the need to use information and communication technology is obvious. The application ranges from basic technologies for carrying out daily management tasks to comprehensive support for difficult decisions using special software systems. In particular, the use of methods from the field of *artificial intelligence* is an increasing trend in practical applications. This area includes the concept of *agents*. An agent can be described as a software entity or a robot (hardware), that is able to perceive its environment and to act upon that environment autonomously [17]. Hence, an agent can be for instance a part of an automatic workflow or an individual representative of a real-world person. Taking individual stakeholders into account as well as the need for flexibility as described in Sect. 1, the usage of methods from the field of *multiagent systems* (MAS) and *agent-based simulation* (ABS) seems promising. They can be used in many different ways. For example, a distributed software system can be used to support automatic coordination and group decisions of real-world participants in their operational activities. In particular, multiagent technology is known for offering flexible solutions and adaptive IT systems [9]. Furthermore, assistance systems with ABS components are able to provide decision support based on the execution of simulation runs, which try to imitate the behavior of the real system. Evaluating various ideas on an artificial system as a simulation model of the real world can be less expensive and time-consuming. The use of multiple agents as a modeling paradigm to build artificial societies or social systems is a unique way of testing theories for many application domains [12]. Beside that, simulation can also be used to evaluate the functionality of a developed MAS by placing the system in a simulated environment. The following description from Wooldridge is helpful for classifying the terms: "A multiagent system is one that consists of a number of agents, which *interact* with one another, typically by exchanging messages through some computer network infrastructure." [21, p. 5]. Thus, an agent-based simulation can be seen as an MAS as well. Despite that, in the following we will use the term MAS to describe a distributed software system and distinguish it from a software system which make use of an agent-based simulation.

The development of both MAS and ABS can be observed in relation to the domain of home health care. The term *home health care* refers to "the provision of healthcare services to people of any age at home or in other noninstitutional settings" [5, p. 9]. To distinguish skilled medical services and non-skilled services, like personal care routines, household maintenance, and social service, the latter is described using the term *home care*, while *home health care* includes medical treatments, nursing services, and physical therapies [16]. To support management in both sectors, various research areas are working on innovative methods. For instance, in *operations research* scientists work on the optimization of daily routing and scheduling for HHC services [6]. Here, engineers, social scientists, and computer scientists, among others, are working on similar problems. To reduce the coverage of the entire range of operational management tasks, the following sections focus on supporting the HHC service provider's planning and resource scheduling. This refers in particular to scheduling processes of employees, i.e., which employee takes on which tasks at which point in time.

3   Review Methodology

As mentioned in the previous section, various approaches exist that apply MAS and ABS in home health care. In order to investigate how and to what extent existing approaches contribute to the operational management of HHC systems, applicable approaches must be identified and analyzed. The conducting of a systematic literature review seems reasonable. For this purpose, search criteria must be defined and applied using a methodologically sound procedure. In this section, both key features for the review and corresponding methodology are presented.

3.1   Literature Search

The identification of relevant approaches, which will be analyzed with the use of the presented key features, was started in March 2018 as a systematic literature search. To this end, a *snowballing procedure* was chosen: The reference list of a scientific paper (*backward snowballing*) as well as the list of papers citing this paper (*forward snowballing*) is used for identifying new relevant papers to examine and the references from as well as to these selected papers are also used in further iterations [20]. The application process can be summarized in the following listing.

1. We generated a literature start set by selecting relevant papers with the help of a web search engine.
2. The reference lists of this start set were used to find further relevant papers, so a second literature set was created.
3. Further, the list of papers citing elements of this start set were examined which generates a third literature set.
4. The references of the second set were examined and no further relevant papers could be found.

5. The list of papers citing papers of the second set were examined and no further relevant papers could be found, too.
6. Thereafter, references of the third set were analyzed and no further relevant papers could be found.
7. Papers citing papers of the third set were also examined and no further relevant papers could be found.
8. All papers found were analyzed in detail and finally 11 contributions can be presented as the result.

The publication dates of the identified approaches range from 2006 to 2017, but in the literature search no temporal restrictions were applied as exclusion criteria. Since background-related biases are possible, it should be mentioned that the education and experience of the authors focus on the field of design-oriented information systems research and artificial intelligence.

As a first step, a web search engine is chosen for the generation of the literature start set. Despite the risk of grey literature *Google Scholar* was chosen because of an absence of knowledge of relevant databases for the considered domain and also because of the fact that the web search engine was only used to create the literature start set. To achieve a small number of iterations in the snowballing procedure, multiple keywords were combined in a search string to generate a suitable literature start set which comprises a high number of papers containing relevant information. The search string used in the web search engine is presented in Table 1. The string contains three groups of keywords, separated by the use of brackets. Each group refers to a domain, which should be represented in a search result. To increase the probability that the content of a search result is a combination of contents of all three domains, the groups are concatenated with logical conjunctions. The first group of keywords specifies the considered operational management in terms of the HHC provider's planning and resource scheduling. The second group specifies the domain of home health care. The third group focuses on the use of the concept of software agents as described in Sect. 2. The use of quotation marks defines a string-based search. For example, any document containing the character string "plan" will be part of the result set, like documents containing the word "planning" or "planner". Due to different spellings, several alternatives are concatenated with logical disjunctions. Furthermore, disjunctions are used for different keywords which describe the same domain.

Table 1. String used in the search engine for generating literature start set.

Domain	String
Planning and Scheduling	*("scheduling" | "roster" | "plan")*
HHC	*("home-care" | "home care" | "home health-care" | "home healthcare"* *| "home health care" | "home health nursing" | "caregiver"* *| "caregiving" | "long-term care" | "long term care")*
MAS	*("multiagent" | "multi-agent" | "agent-based" | "agent based")*

As the inclusion criterion for the snowballing procedure, each domain listed in Table 1 should be represented such that one contribution addresses the application of an agent-based approach that in some way support planning and scheduling in operational HHC management. 16 scientific papers were selected by examining 200 search results (20 result pages) as a result of the usage of the explicated search string in order to create the literature start set. All relevant papers of this start set were found in the first one hundred results. Following the backward snowballing procedure, the reference lists of those 16 papers were evaluated and a second literature set was generated containing two scientific papers. After another iteration, no further publications were found. Following the forward snowball procedure, papers citing a paper of the start set were examined and a third set was created containing five papers. Further, the list of publications citing one of this set was examined and no further papers were found. The remaining iterations, forward procedure with the second set and backward procedure with the third set, neither generated new papers. Afterwards, all 23 papers were analyzed in detail to find only contributions which support operational HHC management in terms of planning and scheduling comprising the conception of a multiagent system or an agent-based simulation. For example, pure literature studies were sorted out. Finally, 11 contributions can be presented as a result. Further iterations of the snowball procedure have already been counteracted by finding useful results with the web search engine in the first step. In addition, due to the application of a comprehensive search string, several papers within the first literature set have mutual references.

3.2 Key Features

To analyze the suitability of the identified approaches, different perspectives of the scientific process must be considered. Before the respective content is presented in the next section, the categorization and the usage of the review key features are outlined. The key features can be divided in *concept, implementation,* and *evaluation*. As a first step, the *concept* is examined to determine how and to which purpose the agent-based system is utilized. Further, the practical *implementation* as well as the *evaluation* of the system are investigated. While the implementation focuses on the availability of software and hardware systems, the evaluation makes sure that the developed concept is applicable in the field.

Five key features are related to the concept. Beside a brief description of the *approach*'s main ideas, the target group of users is identified. In this regard, the *outcome* or *product* is described that is provided to the user. Moreover, methodical limitations and focus of the considered approach are characterized by the key features *spatial aspects, goals and constraints,* and *agents*. The latter designates the agents, which are identified in the approach. The feature *goals and constraints* comprises the targeted performance measures as well as restrictions of the parameter or solution space. The feature *spatial aspects* determines the consideration of any geographic related entities or factors in the model, such as distance computations, traffic predictions, map data, and regional specifications or restrictions.

After taking a conceptual perspective, the provided implementation is analyzed. When implementing a MAS or an ABS, the use of an existing modeling and simulation (*software*) framework is feasible. By this means, common functionalities are provided, which improves the re-usability of the implemented concept. Here, a differentiation has to be made between free-to-use and commercial frameworks. This is directly related to the key feature *accessibility*, which describes whether or not the implementation is available for further use in terms of the used licensing model as well as the provision, e.g., in a public repository. Furthermore, the *interactivity* of the implemented approaches might vary. While some approaches do not allow for real-time interaction, others are equipped with interfaces, which enable the interaction with one or multiple users and also between the involved users.

In the evaluation perspective, the implemented concept is practically applied to home health care scenarios. In terms of MAS or ABS, the evaluation commonly consists of simulation experiments that are conducted as part of a study. This includes design, execution, and analysis of simulation experiments. The *design of experiments* comprises techniques for the identification of relevant experiments (design points; DP) as well as the systematic limitation of the considered parameter space. For stochastic models, the estimation of the required number of replications (sample size; N) is another important task. In addition, input data is required for the definition of the simulated scenario. The key feature *input data* distinguishes between synthetic and real-world data and gives background information like geographical affiliation. Output data that is generated during the execution of the model must be analyzed to draw conclusions about the observed behavior of the system. Based on this, the key feature *output data analysis* describes what means are applied and what efforts are made for assessing statistical significance.

4   Approaches for Supporting Planning and Scheduling

The goal of this work is to survey existing approaches that make use of ABS or MAS to support operational HHC management in terms of planning and resource scheduling. With respect to the conducting of a systematic literature study, the applied research methodology as well as the key features for the assessment of the surveyed approaches were presented in the previous section. As a result of the execution of the literature study, 11 relevant approaches for home health care management were identified. In this section, a comprehensive overview on as well as a comparison of these approaches is presented, which allows for the identification of shortcomings (cf. Tables 2, 3, and 4).

To judge whether and to what extent each of the specified key features (cf. Sect. 3) is satisfied by the approaches, only evidence is used that is directly provided by the scientific publication in which the approach is proposed. Accordingly, in case specific aspects of the system are not discussed in the publication, it is assumed that the approach is not capable of fulfilling the respective key feature. The same applies for ambiguous descriptions or assertions regarding

Table 2. Overview of the concepts (part 1) of the surveyed approaches.

	Approach	User	Outcome/product
Castelnovo et al. [1]	ABS of home care organization model	HHC service provider	Framework to control the home care processes at an operational level
Itabashi et al. [8]	MAS for negotiation of care schedules	HHC service provider	Communication platform for caregivers and patients, care schedule
López-Santana et al. [10]	Multiagent approach using mixed integer programming model	HHC service provider	Communication platform, scheduling and routing for caregivers
Marcon et al. [11]	Global optimizer and ABS of caregiver behavior to solve routing problems	HHC service provider	System for solving scheduling and routing problem in dynamic context
Mohammadi & Eneyo [13]	Sweep-coverage for efficient monitoring of patients by means of a MAS	HHC service provider	Information management system, solving of scheduling and routing problem
Mutingi & Mbohwa [14]	MAS with satisficing heuristic for staff scheduling	HHC service provider	Theoretical framework for staff scheduling and task assignment
Stojanova et al. [18]	Scheduling algorithm and ABS	HHC service provider	Support system for generation and analysis of staff schedules
Widmer & Premm [19]	MAS for negotiation of caregiving resources using double auctions	HHC service provider	Agent-based decision support system for allocation of resources
Xie et al. [22]	MAS for negotiation between home health agency and practitioners	Home health agency	Iterative bidding framework as a decentralized decision making tool
Xie & Wang [23]	ABS for evaluation of schedules generated by repair algorithm	HHC service provider	System for generating and evaluating schedules
Zarour et al. [25]	MAS/agent-based architecture	Set of HHC partners	Platform for communication and cooperation

functionalities of the approaches. To avoid misinterpretations, the assessment which is presented in this section is not based on assumptions in terms of interpretations of text passages. Instead, the wording of the authors is adopted for the description of the approaches. As the terminology that is used for describing the surveyed approaches is not unified, ambiguousness and terminological inconsistencies might occur in the following discussion of the contributions.

Table 3. Overview of the concepts (part 2) of the surveyed approaches.

	Spatial aspects	Goals and constraints	Agents
[1]	-	-	Assumption: participants of a proposed home care reference model
[8]	-	G: MIN total cost of service; C: skills, date/time interval	interface, schedule, helper
[10]	Arrival/departure location, static travel times, multi-depot	G: MIN travel time, MIN delay arrival time; C: skills, locality, priority	Patients, organizer, coordinator, caregiver
[11]	Random events (e.g., traffic jams and road accidents)	G: agents' decision rule (e.g., MIN travel or waiting time); C: unspecified	Patient, caregiver
[13]	Distance from service provider's facility to patient's location	G: MIN no. of therapists; C: location of patients and therapists	Patient, therapist, hospital
[14]	-	G: MIN schedule cost, MAX patient/worker satisfaction; C: tasks, preferences	Manager, patient, nurse, supervisor, resource, scheduler
[18]	-	G: MIN processing time; C: servicing time	Patient, caregiver
[19]	-	G: MAX social welfare; C: time/priority for service, skills, valuation of patient	Patient, caregiver, auctioneer
[22]	-	G: MIN service costs; C: time, skill set, preferences	-
[23]	GIS map as operative environment in simulation	G: MIN service costs; C: practitioner's availability/eligibility, visit time	Practitioner, healthcare agency
[25]	-	-	Patient, broker, doctor, and each cooperation partner

The framework proposed by **Castelnovo, Matta, Tolio, Saita, and De Conno** [1] consists of an ABS of the interactions between different actors that are involved in home care processes. In this regard, the authors make use of the *contract net protocol* to model task distribution between the agents. The goal of the model is to enable patients to stay at home instead of being forced to stay in professional care facilities in case this is not medically necessary. To give a better understanding of possible involved actors, the authors proposed a reference model for the home care domain. The presented approach is implemented in *Arena* and evaluated in a case study of a *Palliative Home Care Program* from Italy.

Itabashi, Chiba, Takahashi, and Kato [8] presented a more comprehensible approach using MAS for the negotiation of care schedules. Equipping caregivers and patients with PDA devices enables the dynamic request of care services as well as the real-time confirmation or rejection of resulting care schedules. The approach aims at minimizing the overall costs of service as care schedules can be adjusted to efficiently take current care requests into account.

In this negotiation process, individual skills of the caregivers as well as date and time preferences of the patients are taken into account. The authors used *JADE* to implement the approach, yet, only presented a synthetic example request to demonstrate its feasibility.

López-Santana, Espejo-Díaz, and Méndez-Giraldo [10] make use of a multi-objective mixed integer programming model to enable scheduling and routing of caregivers in home health care. To consider driveways in the routing and to minimize travel times and delays, departure and arrival locations of the caregivers are specified. However, the presented approach is limited to a single geographical area and travel times are assumed to be static, i.e., not influenced by road closures or traffic-related delays. The proposed platform works well for small numbers of patients (less than 15) but requires heuristics for the calculation of larger amounts of patients. Like the previously introduced approach, the implementation is based on *JADE*. To this end, the authors presented four scenarios with four different parametrizations of the model to illustrate the variation of waiting times.

Of the analyzed approaches, the system presented by **Marcon, Chaabane, Sallez, Bonte, and Trentesaux.** [11] provides the most sophisticated and realistic routing. The combination of a global optimizer with a simulation of individual caregiver decision behavior using MAS allows for the agents' perception of random spatial events such as traffic jams to minimize travel or waiting times. By this means, new requests can also be considered by the system and included in the scheduling and routing process. Constraints that must be considered during the scheduling and routing are unspecified and provided by mixed integer linear programming (MILP) or heuristics. For the implementation of the system, *NetLogo* were used and a comprehensive evaluation is provided. The authors presented two case studies which were derived from French HHC providers and for each case study 500 working days were simulated. As the proposed model consists of stochastic components, the authors executed 100 replications of each parametrization of the model. Finally, they analyzed the performance of the proposed system according to five properties: efficiency, pertinence, scalability, robustness, and implementability.

In the approach presented by **Mohammadi and Eneyo** [13], the scheduling and routing problem is solved by a central unit and by applying sweep-coverage mechanisms. To this end, the authors goal was not the minimization of travel times but the reduction of the required number of therapists. To demonstrate the feasibility of the proposed algorithm, the authors used a *MATLAB* implementation to execute two scenarios each consisting of ten different parametrizations of the model. To take stochastic uncertainties into account, each simulation run was replicated 100 times.

In contrast to other approaches which aim at optimizing HHC scheduling, the architecture proposed by **Mutingi and Mbohwa** [14] makes use of a satisficing heuristic. Here, a schedule that is acceptable for all caregivers is generated based on specific thresholds. To this end, an acceptable schedule is not necessarily optimal. Still, the authors aimed at minimizing scheduling costs while maximizing both patient and worker satisfaction. Of all the analyzed approaches, this one consists of the most agent types. Besides the types manager, nurse, and patient, the authors defined resource, supervisor, and scheduler agents to accomplish multi-objective decision making. The approach was published in 2013. In 2015, the authors applied the approach to decision making for drug delivery in home care services [15].

Stojanova, Stojkovic, Kocaleva, and Koceski [18] focused on scheduling and did not address the routing problem. The authors illustrated analogies between job shop scheduling in logistics and the scheduling of caregivers and elderly people. In the presented ABS, the individuals from both groups are modeled as individual agents which enables communication between the groups. Unfortunately, the resulting simulation is only presented briefly in the paper such that the implemented mechanics remain mostly unclear. *AnyLogic* was used for the implementation of the model, however, experiments or generated results are not presented.

The decision support system proposed by **Widmer and Premm** [19] makes use of an auction-based protocol (double-auctions) to achieve an optimal allocation of caregivers to dementia patients. By this means, they aimed at maximizing social welfare by taking the time required for each service, the skills of each caregiver, service priorities, and valuations of the patients into account. The specification and justification of the proposed auction protocol is the main contribution of the paper. In this regard, a software architecture as well as dementia-specific requirements are introduced. Unlike other contributions that use simulation for their evaluation, the authors presented a scenario-based evaluation to demonstrate the submission of bids as would take place during an auction. The prototype is developed using only the *Java Development Kit* (JDK) and without a dedicated software framework for agent-based approaches.

Xie, Sharath, and Wang [22] presented an MAS framework that implements an iterative bidding procedure for the negotiation of HHC schedules. The parties that are involved in this negotiation process are just the home health agency and the caregivers, leaving out the patients. As the routing of the caregivers is not the primary goal of the presented system, spatial aspects such as traffic or street maps are not considered. The optimization goal which is pursued by this approach is related to the minimization of service costs. To achieve this, time windows, skill sets of caregivers, and preferences of clients are considered. Even though the authors do not present an implementation of the model, they provide experimental results and compared them to the optimal problem solution generated by means of the optimization software *ILOG CPLEX*. For this purpose, eight scenarios were defined each of which is replicated ten times.

Two years after their publication in 2015, two of the authors from the previously presented work proposed another scheduling approach for home health care. As the approaches differ considerably, the system presented by **Xie and Wang** [23] is discussed as well. Unlike the previous publication, the authors proposed an ABS for generating and evaluating HHC schedules using a repair algorithm. Moreover, a spatial aspect was added, so a GIS map serves as operative environment in the simulation. For the implementation, the authors used

AnyLogic and demonstrate the feasibility of the approach based on ten repair runs. As no information on the chosen scenario is given, it must be assumed that the data basis was generated synthetically.

Table 4. Overview of implementation, experimentation, and domain of the surveyed approaches.

	Software	Interactivity	Design of experiment	Input data	Output data analysis	Domain
[1]	*Arena*	-	Sensitivity analysis	1 case study (palliative home care provider in Milan, Italy)	Average values of a performance measure (waiting time)	Palliative Home Care
[8]	*JADE*	Participants reject/accept proposed schedules	-	1 example of single request (synthetic data)	-	HHC
[10]	*JADE*	Allows for new requests during run time	DP = 16, N = 1, deterministic/ stochastic model (unclear)	4 Scenarios (synthetic data)	Average values of a performance measure	HHC
[11]	*NetLogo*	Real-time request of availability of patients	2 simulations of 500 working days, stochastic model, N = 100 for each decision rule	2 case studies (synthetic data, inspired from classical types of French HHC providers)	Statistical significance (confidence interval), evaluation of efficiency, pertinence, scalability, robustness, and implementability	HHC
[13]	*MATLAB*	Assumption: appointments can be made by patients	DP = 20, N = 100, stochastic model	2 scenarios (synthetic data)	Average values of a performance measure	HHC
[14]	-	Update of preferences and management goals	-	-	-	HHC
[18]	*AnyLogic*	-	-	-	-	HHC
[19]	*JDK*	Caregivers and patients submit bids to an auctioneer	-	1 scenario (unknown data source)	-	Dementia (Home) Care
[22]	-	-	Comparison to optimal solution of 8 model configurations (DP = 8), N = 10, stochastic model	8 scenarios (synthetic data at realistic scale)	Average values of a performance measure (bidding solution payment)	HHC
[23]	*AnyLogic*	-	10 repair runs	Assumption: synthetic data	Average value of a performance measure (costs)	HHC
[25]	*JADE*	Information exchange, service requests and offers	-	Two examples of coordination (synthetic data)	-	HHC

Zarour, Zarour, and Khalfi [25] provide an agent-based architecture to support coordination and communication for patients and cooperating providers of services regarding the patient's care. Similar to the previously mentioned contributions from Castelnovo et al. and Itabashi et al., the protocol *contract net* was used to enable coordination like scheduling on an abstract level. Furthermore, the authors defined the agent's communication processes and presented an ontology for information exchange in the considered domain. The implementation is based on *JADE* and for evaluation the authors compared their information system with a similar agent-based support system. Because of missing information regarding the input data of the scenario for comparison, it must be assumed that they used synthetic data. A part of this work (presented in 2010) was already published in 2008 by two of the authors [24].

Beside these 11 approaches, which were selected in the literature review as described in Sect. 3, the ideas presented by **Fraile Nieto, Rodríguez, Bajo, and Corchado** [7] are worth mentioning. The authors applied an abstract MAS architecture to a home care scenario, where agents can offer and request services from other participants. This can be conceived as a part of a management solution. Because of a lack of elaboration in the area of resource scheduling, the publication is not part of the tables. The authors only mention that it could be possible to use this architecture for scheduling medical staff.

Similar to the previously mentioned contributions using the *contract net protocol*, the work of **De Causmaecker, Demeester, Berghe, and Verbeke** [3] provides an agent-based scheduling approach including negotiations for personnel exchange respectively task exchange. In the same year (2005), the authors published another paper to this topic and give further information regarding an implementation and planned experiments [4]. Because both publications are very short, without details, and not linked directly to the domain of home health care, the contribution is not part of the tables here. By looking at an earlier publication from 2004, a connection to the domain of HHC can be established. Here, De Causmaecker et al. [2] analyzed personnel scheduling problems, mention application domains, and propose a classification, where one type of planning refers to home health care.

5   Shortcomings of the Surveyed Approaches

The previous section analyzed the identified contributions with respect to the defined key features. None of the surveyed approaches is satisfactory to support planning and scheduling in operational HHC management regarding flexibility in caregivers' operations, efficiency in the use of resources, and economic viability under present and future conditions.

Shortcomings in the approaches' concepts are mostly related to *outcome*, *spatial aspects*, and *goals*. It can be observed that an *outcome* for the HHC management that is "ready to use" does not exist. Beside theoretical contributions, like frameworks, the publications provide outcomes on a prototype level. Further, *spatial aspects*, such as traffic times or map data, are not sufficiently considered and no real-world traffic data is used. Instead, static travel times are used, not regarded at all, or no distinction in the direction among the nodes is made. Specific optimization goals are pursued in 9 out of 11 publications and only one system allows for the interchangeability of goals. The approach by López-Santana et al. takes cargivers' skills into account and allows for new customer requests during runtime. However, their system focuses only on minimizing travel times of caregivers as well as delays in arrival times at customer locations. Here, generating an optimal solution takes too much time for real-world problems. Similarly, the approach by Xie and Wang only focuses on minimizing service costs and the search for an optimal solution takes too long here as well. The approach by Marcon et al. assigns each caregiver to a set of customers with a corresponding route proposal, which can be adapted later by the caregiver. Following this, a dynamic solution is provided based on a caregiver's local decision. By changing the local decision-making mechanisms, different higher-level objectives can be pursued, e.g., minimizing waiting times. But each caregiver only interacts with his own patients, so interchangeability is not possible. In addition, there is no coordination between caregivers to react on events in order to reach a better joint solution. The remaining surveyed approaches do not provide sufficient dynamic scheduling solutions, but communication platforms and basic coordination solutions.

In the implementation of the surveyed approaches, shortcomings are observed in terms of used *software* and its *accessibility*. Through the use of commercial frameworks, a third-party is included which claims license fees for use. Consequently, a monetary dependency results. Further, a dependency arises in software maintenance and durability. Overall, the applicability of the implementation is strongly limited. Regarding the *accessibility* of the implementations, none of the authors referred to online repositories or websites for downloading the proposed implementations. Because of the inaccessibility of all developed software, the key feature *accessibility* is not part of the tables. Shortcomings in the evaluation of the surveyed approaches arise in all defined key features. First, relevant parts of the parameter space must be identified and systematically investigated. Unfortunately, the *design of experiment* in the publications is mostly on a non-professional level. Second, *input data* in terms of suitable real-world data is not provided sufficiently. Either synthetic data or a brief case study is given. Third, to ensure statistical reliability and the significance of the evaluation results, it is recommended to apply means of *output data analysis*. The greater part of the surveyed approaches uses information about considered performance measures in terms of statistical measurements of central tendencies.

6   Conclusion and Further Work

Operational management processes in terms of planning and scheduling in the domain of home health care systems can be supported by multiagent systems as well as decision support systems using agent-based simulation. This article provides an overview of these approaches. Therefore, we conducted a systematic literature review in which 11 relevant approaches using agent technology were identified. Further, the identified publications were analyzed and shortcomings were detected. The shortcomings comprise aspects of the respective concepts, the provision of implementations and the execution of evaluation processes. In order to cope with an increasing demand in HHC, besides efforts to improve efficiency, additional caregivers must be hired by service providers. However, the availability of skilled caregivers on the labor market is very limited. None of all current agent-based approaches offers a sufficient solution to dealing with a shortage of skilled workers. Furthermore, no learning mechanisms for agents are used to increase efficiency and the handling of the dynamic environment is not sufficient as well. The latter includes coping caregiver outages and delays in operational processes as well as no usage of real-world traffic data.

As the need for management support persists, we are working on a dynamic distributed scheduling solution for the control of operational processes which increases efficiency and improves the use of limited resources to allow for coping with the rising demand.

References

1. Castelnovo, C., Matta, A., Tolio, T., Saita, L., De Conno, F.: A multi agent architecture for home care services. In: Tavakoli, M., Davies, H.T. (eds.) Reforming Health Systems: Analysis and Evidence: Strategic Issues in Health Care Management, pp. 135–151 (2006)
2. De Causmaecker, P., Demeester, P., Berghe, G.V., Verbeke, B.: Analysis of real-world personnel scheduling problems. In: Proceedings of the 5th International Conference on Practice and Theory of Automated Timetabling, Pittsburgh, pp. 183–197 (2004)
3. De Causmaecker, P., Demeester, P., Berghe, G.V., Verbeke, B.: A coordination model for distributed personnel scheduling. In: Proceedings of the 19th Orbel Conference (2005)
4. De Causmaecker, P., Demeester, P., Berghe, G.V., Verbeke, B.: An agent based algorithm for personnel scheduling. In: Models and Algorithms for Planning and Scheduling Problems, Siena, pp. 103–105 (2005)
5. Dieckmann, J.L.: Home health care: a historical perspective and overview. In: Handbook of Home Health Care Administration, pp. 9–26 (2015)
6. Fikar, C., Hirsch, P.: Home health care routing and scheduling: a review. Comput. Oper. Res. **77**, 86–95 (2017)
7. Fraile Nieto, J.A., Rodríguez, S., Bajo, J., Corchado, J.M.: The THOMAS architecture: a case study in home care scenarios. In: Workshop on Agreement Technologies (WAT) (2009)
8. Itabashi, G., Chiba, M., Takahashi, K., Kato, Y.: A support system for home care service based on multi-agent system. In: Information, Communications and Signal Processing, pp. 1052–1056. IEEE (2006)
9. Kirn, S.: Flexibility of multiagent systems. In: Kirn, S., Herzog, O., Lockemann, P., Spaniol, O. (eds.) Multiagent Engineering: Theory and Applications in Enterprises, pp. 53–69. Springer, Heidelberg (2006). https://doi.org/10.1007/3-540-32062-8_4

10. López-Santana, E.R., Espejo-Díaz, J.A., Méndez-Giraldo, G.A.: Multi-agent approach for solving the dynamic home health care routing problem. In: Figueroa-García, J.C., López-Santana, E.R., Ferro-Escobar, R. (eds.) WEA 2016. CCIS, vol. 657, pp. 188–200. Springer, Cham (2016). https://doi.org/10.1007/978-3-319-50880-1_17
11. Marcon, E., Chaabane, S., Sallez, Y., Bonte, T., Trentesaux, D.: A multi-agent system based on reactive decision rules for solving the caregiver routing problem in HHC. Simul. Model. Pract. Theory **74**, 134–151 (2017)
12. Michel, F., Ferber, J., Drogoul, A.: Multi-agent systems and simulation: a survey from the agents community's perspective. In: Weyns, D., Uhrmacher, A. (eds.) Multi-Agent Systems: Simulation and Applications, Computational Analysis, Synthesis, and Design of Dynamic Systems, pp. 47–51. CRC Press - Taylor & Francis (2009)
13. Mohammadi, A., Eneyo, E.S.: Home health care: a multi-agent system based approach to appointment scheduling. Int. J. Innov. Res. Technol. **2**(3), 37–46 (2015)
14. Mutingi, M., Mbohwa, C.: A Home healthcare multi-agent system in a multi-objective environment. In: SAIIE25 Proceedings, pp. 636/1–8 (2013)
15. Mutingi, M., Mbohwa, C.: Developing multi-agent systems for mhealth drug delivery. In: Adibi, S. (ed.) Mobile Health. SSB, vol. 5, pp. 671–683. Springer, Cham (2015). https://doi.org/10.1007/978-3-319-12817-7_29
16. Prieto, E.: Home Health Care Provider: A Guide to Essential Skills. Springer, Heidelberg (2008)
17. Russell, S., Norvig, P.: Artificial Intelligence: A Modern Approach. Prentice Hall Series in Artificial Intelligence. Prentice Hall, Upper Saddle River (2010)
18. Stojanova, A., Stojkovic, N., Kocaleva, M., Koceski, S.: Agent-based Solution of Caregiver Scheduling Problem in Home-Care Context, pp. 132–135 (2017)
19. Widmer, T., Premm, M.: Agent-based decision support for allocating caregiving resources in a dementia scenario. In: Müller, J.P., Ketter, W., Kaminka, G., Wagner, G., Bulling, N. (eds.) MATES 2015. LNCS (LNAI), vol. 9433, pp. 233–248. Springer, Cham (2015). https://doi.org/10.1007/978-3-319-27343-3_13
20. Wohlin, C.: Guidelines for snowballing in systematic literature studies and a replication in software engineering. In: 18th Conference on Evaluation and Assessment in Software Engineering, pp. 38:1–38:10. EASE 2014. ACM (2014)
21. Wooldridge, M.: An Introduction to MultiAgent Systems. Wiley, Hoboken (2009)
22. Xie, Z., Sharath, N., Wang, C.: A game theory based resource scheduling model for cost reduction in home health care, pp. 1800–1804. IEEE (2015)
23. Xie, Z., Wang, C.: A periodic repair algorithm for dynamic scheduling in home health care using agent-based model, pp. 245–250. IEEE (2017)
24. Zarour, K., Zarour, N.: An agent-based architecture for a cooperation information system supporting the homecare. In: 2008 International Workshop on Advanced Information Systems for Enterprises (IWAISE 2008), pp. 63–69. IEEE (2008)
25. Zarour, K., Zarour, N., Khalfi, S.: Inter-agent interaction in medical information system. J. Theor. Appl. Inf. Technol. **11**, 130–142 (2010)

Ethical Surveillance: Applying Deep Learning and Contextual Awareness for the Benefit of Persons Living with Dementia

Steve Williams[1] , J. Mark Ware[1] , and Berndt Müller[2]

[1] Faculty of Computing, Engineering and Science,
University of South Wales, Pontypridd, UK
{steve.williamsl,mark.ware}@southwales.ac.uk
[2] Department of Computer Science, Swansea University, Swansea, UK
berndt.muller@swansea.ac.uk

Abstract. A significant proportion of the population has become used to sharing private information on the internet with their friends. This information can leak throughout their social network and the extent that personal information propagates can depend on the privacy policy of large corporations. In an era of artificial intelligence, data mining, and cloud computing, is it necessary to share personal information with unidentified people? Our research shows that deep learning is possible using relatively low capacity computing. When applied, this demonstrates promising results in spatio-temporal positioning of subjects, in prediction of movement, and assessment of contextual risk. A *private* surveillance system is particularly suitable in the care of those who may be considered vulnerable.

Keywords: Privacy · Deep learning · Assisted-living · Mobile computing · Ethics · mHeath · Wearable health · Dementia · Safer walking · GPS · LSTM · RNN

1 Background

Advancements in mobile devices that can be worn and carried, their interconnectivity, and the improvement of artificially intelligent tools provide a significant opportunity to assist in the care of the aged. In accordance with a human right to private life, we have examined methods to keep tracking information private unless there is a moral argument, such as risk to the person being monitored, that justifies a breach in privacy. In this scenario, safety is paramount and in the interests of beneficence and non-maleficence an ethical policy in terms of design is employed, which defines that personal information is precious and should therefore not be shared on the internet.

Dementia is a debilitating condition that is growing with the aging society. Continuance with life in the community is encouraged, since social interaction and physical activity stimulates a healthy mental state in the person with symptoms (PwS) along with the family carer. We seek bespoke artificially intelligent solutions for these persons living with dementia (PlwD) who wish to preserve independence of the PwS.

F. Koch et al. (Eds.): AIH 2018, LNAI 11326, pp. 29–47, 2019.
https://doi.org/10.1007/978-3-030-12738-1_3

Initial system infrastructure and findings are published in [1], the suitability of a mobile computer technology in tracking PwS and ethical aspects are previously outlined in [2]. The work described here contributes to the ethical debate regarding the question at which point information gathered when monitoring a PwS should be shared. We investigate a technological solution that keeps data private until a threshold of risk is reached. AI is used to learn what is 'normal' for a person (based on individual habits), various metrics are then used in the decision making to change the default private state.

To this end, a monitoring system is designed that requires the PwS to carry a mobile phone and wear a fitness tracker. It is understood that some may not be comfortable with this and it is anticipated that the mobile technology component will ultimately be integrated in a single wearable device. This technology can be particularly useful for patients who have early-onset dementia, i.e., those of working age and therefore more likely to be used to carrying a phone or wearing a smart-watch.

2   The Problem

The onset of dementia has a profound effect on the PwS and the wider family unit. Diagnosis can bring with it a loss of role function, uncertainty about the future, fear of being a burden, and reduced mobility that can lead to social isolation [3]. The objective of this study is to create an 'electronic safety net' that can provide peace of mind to the carer, while preserving the rights and independence of the PwS. A key aim of the project is to delay residential care.

2.1   Dementia

Dementia is caused by several diseases of the brain. There is a wide spectrum of symptoms, some of which may manifest in a propensity to walk independently at inappropriate times [4]. Literature indicates that this can lead to premature mortality [4–6]. Actions to mitigate this risk can lead to increased dependence, to curtailment of social activities, and reduction in quality of life [7]. Elopement episodes are a major reason for nursing home admission [8]. A study in Finland reports that the latter may be delayed, using assistive technology, by an average of eight months [9].

2.2   Privacy

Online data privacy divides opinion. Many elect to share very varied information about their lives publicly on the internet, but this is not always a conscious decision – Terms and conditions regarding data sharing tend to be ignored by many users as they install applications and use online services. Nevertheless, consent given in this way is often referred to as informed when the potential for data propagation is mentioned in the supplied information, even though this information is rarely considered thoroughly.

Leaks of private information have recently been in the news headlines. Data stored on the internet, e.g. by cloud services, is often assumed to be safe, but human intervention and inadequate security measures allow breaches [10]. Advocates of privacy treat personal information very differently and avoid sharing their information with people or organisations. This attitude is supported by cyber-security activists, e.g. in a report of vulnerabilities leading to 91 exploits of tracking service providers in January 2018 [11]. In the case of care for persons who may be considered vulnerable it seems ethically correct that a strict data protectionist policy should be the default.

2.3   A Human Rights-Based Approach

The World Health Organisation (WHO) advocates a human rights-based approach for PlwD [12]. In our study, almost two years of personal data was collected. This included location (derived using GPS and nearby Wi-Fi nodes), activity recognition, indoor movement, and logs of heart rate, steps, and sleep patterns. This kind of monitoring undoubtedly has the potential to invade a person's right to private life. The tracking was described by the subject as a big-brother bad dream. On reflection, the level of 'invasion' depends on who has access to the data.

3   Machine Learning (ML)

The aim is for an algorithm to learn human mobility patterns of an individual, and to assess the perceived risk against the learnt normality that is deemed to be 'safe'. A measure of risk is used to determine the level of protection required on the personal data collected. To protect privacy, propagation of this information is restricted to the secure home network. No interaction with the wearable or phone is required of the PwS.

To improve potential accessibility to many users in the long run, the equipment used in a working prototype is a standard smart-phone and a home-based 'hub', which is a credit-card sized computer with limited resources, such as a Quad-Core 1.2 GHz CPU and 1 GB RAM. Networking between the two in 'monitoring' mode is via on-board Bluetooth and Wi-Fi only while at home.

Unconventional Deep Learning: Deep learning (DL) discovers intricate patterns in large datasets by using multiple processing layers to learn representations of data [13]. Sequential and parallel information is processed in a cyclical (recurrent) fashion by modifying internal weightings of input signals to produce an expected output signal [14, 15]. The hardware platform described may seem restrictive for a DL task in an age where we are used to resources being server based and ubiquity being the norm. Convention says that DL requires large computing capacity, but this is not available for the present use case. Long Short-Term Memory (LSTM) networks [16] are a type of Recurrent Neural Network suitable for learning and predicting sequential patterns in timelines. Using accelerometers, as commonly found in modern mobile devices, LSTM are deployed in human activity recognition (HAR). X-Y-Z accelerometer readings are interpreted over a defined time-period and then compared to those taken in a laboratory to determine probability that a categorised activity is taking place [17]. We have assimilated this using GPS sensor data. A dataset suitable for learning using an LSTM neural network was developed, and the resultant tensor was deployed to an Android device to calculate the probability of being on a learnt trajectory or otherwise.

The novel concept that surveillance need not be *invasive* is introduced. There is a host of literature relating to HAR [18], there are indoor monitoring studies with AI, e.g. [19], and studies of wandering trajectories, e.g. [20]. None of these describe categorisation of the normal movements of a person together with discrete monitoring that keeps information private until anomalies are found.

3.1 ML Methodology

Data: GPS data is collected from one subject using a standard HTC-10 smartphone used solely for that purpose. Considerable data preparation is required using the minute-by-minute location coordinates. Data is first compartmentalised based on total movement to date (*tm*). This is then divided by an increment (*i*) giving sub-divisions as shown in Fig. 1 with $i = 20$.

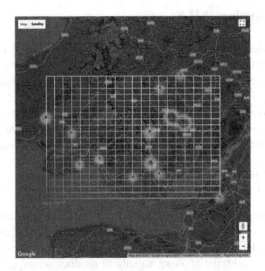

Fig. 1. Boundaries of the extent of total movement for 3 months, i = 20. Map: © Google.

To optimise computation time, daily data is reduced to only the proportion that represents movement.

Categorisation. Points within each segment (or compartment) are assessed for each trajectory and each segment's points are compared using a kd-tree-based nearest-neighbour algorithm [21]. The degree of similarity is assessed giving a percentage and a threshold provides a similarity decision. There is difficulty in some trajectories where, for example, topographical, atmospheric, or networking issues used in test data collection lead to sparse and noisy data. Sparse data was dealt with using 1d-univariate interpolation [22]. This is particularly important in the early days of training where there are few trajectories to compare. Noisy data is essentially ignored at this time by adding a tolerance to the similarity decision just described which is explained in more detail below. The result of the comparison algorithm is a segment chain (string) for trajectories with 1 or 0 signifying a match in each square (Fig. 2).

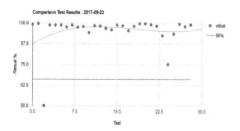

Fig. 2. A successful match of two segment chains. 29 segments, $i = 10$.

Fig. 3. Interpolation used to deal with sparse data causing accuracy issues. Map data: © Google.

Categorisation by comparison of trajectory segment chains by only comparing matched segments significantly reduces the computational capacity required in terms of processing and memory. If a match is found, interpolated point data is added to a master repository with which future comparisons are made. An encoded polyline [23] reduces database size requirements and allows trajectories to be stored as entities. In time, the necessity for interpolation is reduced as the repository trajectory density increases.

As seen in Fig. 3, interpolation may cause significant deviation from the route that is travelled, e.g. by cutting corners and using roundabouts, but this level of granulation is considered satisfactory as a 'zone of safe movement' is maintained. Matching segments rely on a nearest neighbour tolerance (*nnT*) and merging with subsequent trajectories eventually creates a dense category master that is used to define this zone.

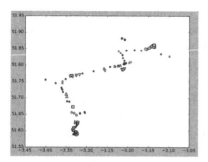

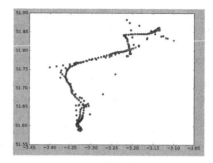

Fig. 4. (a) A comparison tolerance *nnT* leads improved matches while ignoring noise. (b) Interpolated points are merged to create a dense category master.

nnT set at 0.005 in decimal degrees, equating to just over 500 m, is used in the experiments. This tolerance can be linked to *tm* in further work as the extent of movement defines the granularity required within the movement space. The resultant categories develop into a densely populated polyline seen in Fig. 4b. All movement within a data collection period are matched with destinations recognised in the initial cluster analysis.

In addition to our collected data, the comparison algorithm was tested using seven users' data from the Geolife (GL) dataset [24]. This contains better quality GPS trajectories and includes higher variance in modes of travel. With *nnT* applied to nearest neighbour algorithm it is observed that small deviations from a route are not a significant problem. As can be seen in Fig. 5, four separate tracks converge on a destination and in the extent of this day's movement all points are within one segment.

Noise, detours and differing distances included in two tracks taking Route 1 and Route 2 in Fig. 6, both arrive at the same place E^1 and C^2. *nnT* allows for the eventuality of C^1 and D^1 not matching Route 2. Adding both to the master increases the possibility that subsequent trajectories match by widening the dataset.

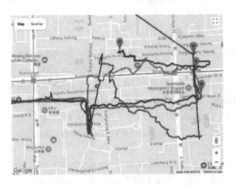

Fig. 5. Detours and converging paths are handled using *nnT* and segment comparison. Map: © Google Maps.

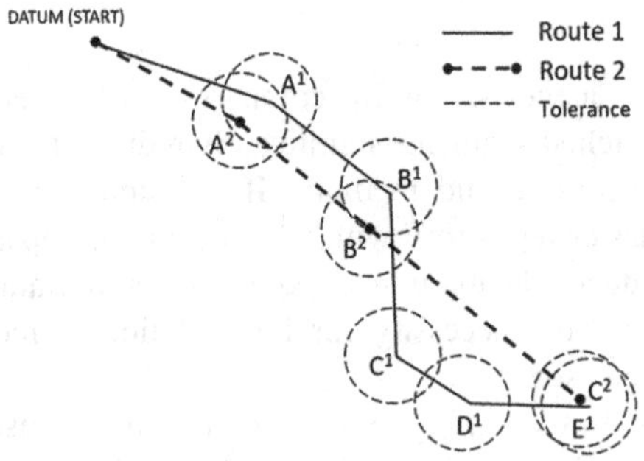

Fig. 6. Widening the category master by allowing a nearest neighbour tolerance.

Bearing. Some GL users' data highlighted the difficulty of recognising direction of travel in that only one-way trajectories are recorded. Experimentation with inclusion of direction of travel gave complex results, consequently movement is treated as omni-directional; the category master is essentially an amalgamation of history on that route.

Time Factor. This is an important consideration in the study scenario, but the likelihood of a person travelling a recognised trajectory at the same time is low so prediction of this is not required. There are detours from a route, the method of travel may change, there may be traffic. These factors all have a significant impact on spatio-temporal data and following extensive experimentation, it is concluded that data-point true timestamps cause confusion. Instead, each category master is indexed sequentially.

Predictability. Major studies in human mobility patterns find that there is a high degree of temporal and spatial regularity [25]. In the datasets investigated, this study concurs; the number of places travelled to is surprisingly low. The three most regularly visited destinations are selected for demonstration; these are travel to University (south), to social visits (west) and to a supermarket (east) seen in Fig. 7.

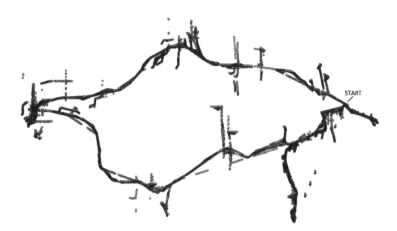

Fig. 7. Three categories of travel overlaid with noise showing, interpolated; 3×10 k records.

Pre-processing. Category masters are exported and the number of records per category is equalised by interpolating (increased or reduced) to 10,000 records each. Noise is amplified where outliers are interpolated. These outliers will be removed in later versions of the system. The data is stacked and normalised. Train:test split is 80:20.

Machine Learning. Inspiration for this is credited to work using Convolutional Neural Network and LSTM RNN in mobile phone HAR applications. The solution selected for our application is Tensorflow 'BasicLSTMCell' stacked with 'Multi-RNNCell' with 64 hidden units. The neural network is expected to learn geo-spatial data to predict categorisation (of the trajectory) when it is given further blocks.

3.2   ML Results

Experimentation found that the number of time steps set at 10, in blocks of 10 gave an accuracy of 90–97% over 500 epochs in less than 1.5 h (Fig. 8).

Deployment. Using our dataset, the resultant tensor is imported to an Android application that sequentially passes arrays of 10 steps of a test trajectory in a timed fashion. A Tensorflow classifier returns the probability of the array being Category 1, 2 or 3 for the three trained classes. These predictions are logged on the phone (Fig. 9).

Mobile Results
Category 1: Correctly predicted with 98–99% certainty unless trajectories overlap.
Category 2: Correctly predicted with 55–86% certainty.
Category 3: Correctly predicted with 77–90% certainty.

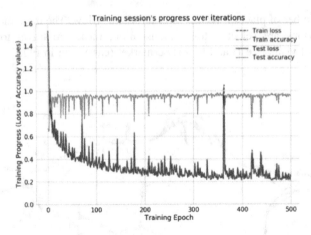

Fig. 8. LSTM training session over 1.4 h. 90–97% accuracy.

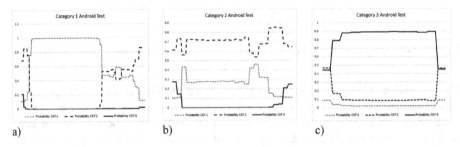

a)   b)   c)

Fig. 9. Android category prediction results: the vertical scale on these graphs range from 0 to 1 where 1 = 100% certainty. (a) Category 1 (dotted line). (b) Category 2 (dashed line). (c) Category 3 (solid line).

The Tensorflow classifier gives reliable prediction of a route being tested in all cases. These are very satisfactory results. Overlap between two categories returning a 50:50 result in the Category 1 test is perfectly acceptable since the routes do overlap.

Public Dataset Results

Training was carried out using matched trajectories in the GL dataset with similarly acceptable results (Fig. 10).

Volunteer Test Results

Six volunteers were recruited from a convenience sample that consented to be tracked by phone and fitness tracker for a period of three months. GPS data was collected by enabling Google Timeline and by configuring their phone accordingly. Places visited, and routes taken are stored on Google servers[1]. At the end of the period data was

---

[1] Note that Google Timeline is only used for data gathering in this initial feasibility study. The full solution uses GPS data stored only locally on the mobile device and processed on the home hub.

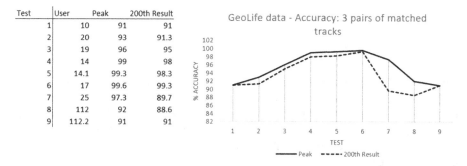

Test	User	Peak	200th Result
1	10	91	91
2	20	93	91.3
3	19	96	95
4	14	99	98
5	14.1	99.3	98.3
6	17	99.6	99.3
7	25	97.3	89.7
8	112	92	88.6
9	112.2	91	91

Fig. 10. LSTM training session for 7 users (in 9 tests) using Geolife dataset.

exported, and a bespoke script was developed to interpret the data. With very few exceptions, it was found that the main locations visited can be classified as attending a place of work or study, going shopping and visiting friends or family.

Machine Learning Results: As previously explained, the six datasets consist of latitude, longitude, and elevation. When subjected to the same neural network, training test results are shown in Table 1.

Table 1. Volunteer machine learning results

Volunteer	Age	Gender	Phone	% accuracy result
1	46	F	iPhone 5s	86
2	23	M	iPhone 6s	84
3	22	F	iPhone 5s	85
4	80	F	Android Galaxy S5	89
5	45	M	Android Galaxy J3	97
6	54	M	Android Galaxy S4	88

Volunteer 3 presented problems in that trajectories overlapped due to the topography of their home address. A revised method that change the way overlapping trajectories are categorised can be used overcome these issues. Volunteer 4 had noticeably more restricted movement and comparably reduced distance travelled causing specific issues of data sparsity. For these reasons both datasets required manual categorisation and matching.

Data Augmentation: Noisy data was dealt with using Google Snap-to-Road [26] and/or TrackMatching [27] and when sparsity occurred, gaps were filled using route finding techniques such as Google Directions API [28] or a variation of Open Street Maps routing [29]. Route finding methods of augmentation cause an element of subjectivity but provided data suitable to test the network.

Deployment Results: The Android simulator gives closely comparable results as that with our own data, for example volunteer 1:

Category 1: Correctly predicted with 42–54% certainty.*
Category 2: Correctly predicted with 97–99% certainty.
Category 3: Correctly predicted with 98–99% certainty.
* routes overlap

It is concluded from this series of tests that the developed machine learning model provides adequate accuracy in the categorisation of routes. The differing data collection methods give a valuable insight into how best to develop training data. The method of transferring learnt information to a mobile phone is particularly interesting as heavy processing can be carried out on a hub, thus preserving the restricted battery resource of the mobile device.

4 Situation Appraisal

Real-time appraisal of the situation of the person being monitored is key to ensuring their well-being. The system is designed with PlwD in mind, so apart from elopement, issues specific to PwS are considered:

Sleep and Dementia. Circadian rhythm disorders can present as an early component of the disease. They have significant impact on patients and caregivers and are a 'major risk factor for early institutionalisation' [30, 31]. Symptoms include sleep disturbances, sun-downing, and agitation. Instances of elopement regularly occur at night. Disturbance in sleep of the PwS has a significant effect on care-givers that can lead to their ill-health [32, 33].

Therefore, monitoring of sleep is highly relevant to this study. Not only should PwS's safety outdoors be monitored, but a metric of well-being should be used to modify system sensitivity. The following section describes the approach and some technicalities of machine learning in this area. In addition, the initial metrics used in appraisal of the contextual situation the person being monitored is in are outlined. Factors such as sleep, and heart rate are here referred to as the 'pre-disposition' of the person. This may be understood as a metric for their well-being.

Data Collection. Although sharing data to a manufacturer's server breaches the complete privacy rule, a FitBit fitness monitor is used in this study for convenience. A dedicated wearable with direct, local data access would allow to preserve privacy and will be used in the final prototype. A FitBit 'Ionic' is one of many devices that are popular with those who wish to monitor, for example, a keep-fit regime. While wrist actigraphy is customarily used in sleep research there is evidence that FitBit devices provide close estimation of total sleep time [34]. Over 2 years of data was collected from one subject using this device and more than three months of data from six volunteers using similar models. It was found that the data collected gives a good representation of actual sleep patterns. A secure authenticated oAuth2.0 API is used to access data from the FitBit servers, yielding daily data, when visualised is shown in Fig. 11.

Machine Learning. Machine-learning techniques have been developed that assess the data, which includes minute by minute heart-rate, steps and sleep records. The

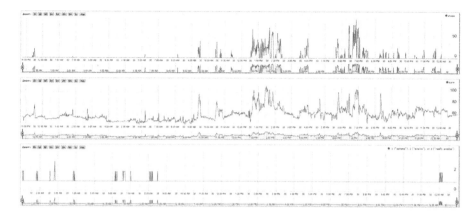

Fig. 11. Graph of steps, heart rate and sleep records for one day.

requirement is that human activity is discretely monitored with automated realisation of trends. The daily situation is then evaluated and compared to what is deemed 'normal' for the individual.

Sleep Period. This is modelled using a Gaussian Distribution in order to give clarity on the expected duration of sleep. Long term changes to averages in a 3-month moving period for example using start, finish and duration of sleep may be used in appraisal (cf. Fig. 12).

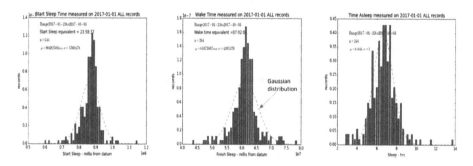

Fig. 12. Gaussian distribution of sleep start, finish and duration for 3 months.

While the subject in this study does not suffer from disruption in diurnal rhythm per se, average sleep per week, and distribution, give an interesting illustration that show variance in the time-period (cf. Fig. 13).

Results can be categorised using 1σ or 2σ, i.e. 68% or 98% of the norm (μ). Waking times can be defined as normal (up to 1σ), early or late (between 1σ and 2σ), and very early or late ($>2\sigma$). When visualised, trends are apparent, there are outliers that represent exceptional occurrences in this period.

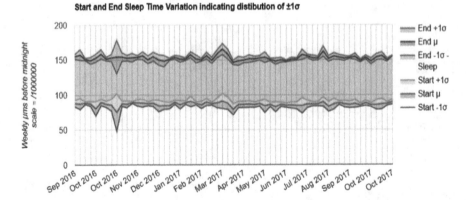

Fig. 13. Trends in weekly sleep time for one year ignoring restlessness.

Activity. The same method was applied to step-count in daytime and night periods. This information is useful in recognising active periods during the night and possible association of these with daytime sedentary periods possibly in correlation with less sleep at night.

Heart Rate While Asleep. It is observed in our dataset that in the day, heart rate closely relates to physical activity such as steps, but while asleep, lack of movement can be used to provide a period in which it is possible to benchmark and provide reliable regression analysis. As illustrated in Fig. 11, minute by minute daily heart rate is collected. When heart-rate while asleep is extrapolated across days and polynomial regression compared, clear differences are evident. Centroids of five sleep periods are analysed. The sleep periods are *start*, *early*, *mid*, *late* and *finish*. In this way, varying sleep periods of different lengths are normalised. Agglomerative clustering with simple Euclidean affinity [35], and k-means cluster analysis [36] are used to give single centroids for each period. Having results for each period makes it possible to visualise clusters (cf. Fig. 14), and conclude μ and σ in any defined period.

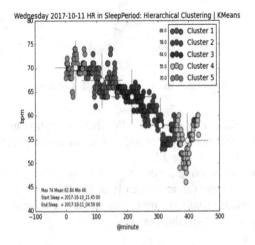

Fig. 14. Five clusters of heart rate readings in one night. $\mu = 63$ bpm.

Centroid data for a date period gives an average of averages (μ). If a limit of, for example, ($\mu + 0.5\sigma$) is applied on data available then results can be categorised as being ordinary (0) or otherwise (1). Currently, only checks for high heart rate are introduced into the training data, but others could be included in future.

Neural Network. A neural network was developed using Keras and TensorFlow. The input layer and 2nd hidden layer uses a rectifier activation function with 6 nodes and 9 inputs. These are maximum, minimum, μ, σ and the five centroid results. A sigmoid activation function is used on the output layer. The classifier function is compiled using the 'adam' optimiser with 'binary_crossentropy' loss, 500 epochs are used in training in batches of 10. The data is split so that 80% is used in training and 20% as test data. Experimentation using just over a year of raw-data were useful in that they were able to recognise the categorisation that had been applied to that data. 94.9% accuracy was measured for 1 year's data in 105 s. Using 1 month of data, 85% accuracy was measured in 24 s, this was then improved to 87% by only supplying centroid measurements i.e. 5 inputs. It was concluded that pre-processing using hierarchical clustering then k-means and applying a fixed rule for categorisation is reliably recognised in a Neural Network; it is possible, for this subject, to predict with 87% accuracy given a month of data. The categorisation of data requires a rule to build training data, but once training has taken place the resultant tensor is deployed to the phone to process daily readings autonomously. The system successfully recognised exceptional heart rate events of the subject.

A fitness tracker that is worn 24 h a day provides an efficient way to collect information for this study. The product used is aesthetically pleasing and if a PwS is used to wearing a smart-watch it should not present a problem in use. Activity and indication of heart-rate levels may provide a useful indicator of well-being of a person at night.

5 Contextual Factors

When coupled with fundamental contextual factors such as time-of-day, distance from home, and weather conditions, contextual risk of being at a location outdoors can be used in decisions regarding preservation of privacy. The following sections summarise factors used in this study.

5.1 Time and Distance Metric

Time-of-day is easily determined on a computer and is an important factor when considering risk. Weightings $w(t)$ of time t and distance from home are used for analysis. A time metric simply uses the hour of day, this is provisionally set as follows:

$$w(t) = \begin{cases} 1 & \text{if } 9{:}00 \leq t < 18{:}00 \\ 2 & \text{if } 6{:}00 \leq t < 9{:}00 \text{ or } 18{:}00 \leq t < 21{:}00 \\ 3 & \text{if } 21{:}00 \leq t \text{ or } t < 6{:}00 \end{cases} \tag{1}$$

with 1 representing low risk. A distance metric $w'(d)$ for distance d is similarly set at:

$$w'(t) = \begin{cases} 1 & \text{if } 0\,\text{km} \le d < 1\,\text{km} \\ 2 & \text{if } 1\,\text{km} \le d < 5\,\text{km} \\ 3 & \text{if } 5\,\text{km} \le d \end{cases} \tag{2}$$

These definitions are notional values for use in the experiments which should initially be decided by the user dyad after considering the questions of *when* and *how far* is normal for the individual PwS.

5.2   Weather Metric

In a similar way, scales of risk can be compiled using weather forecasting applications such as Dark Sky or OpenWeatherMap. The Dark Sky API [37] offers a full collection of meteorological conditions and is used in the study. The locality of the subject is known so forecast data is retrieved for temperature, precipitation and wind in that area. A rudimentary weather metric is defined using a matrix (see Table 2 where, again, 1 corresponds to low risk).

Table 2. Weather metric matrix

Weight	Temperature (°C)	Precipitation (mm/h)	Wind (Beaufort scale)
1	]15..25]	[0..1]	[0..5]
2	]10..15],]25..27]	]1..4]	]5..11]
3	[∞..15],]27..∞]	]4..∞]	]11.. ∞]

Figure 15 illustrates how this matrix can be used to conclude an accumulated weather measurement and how this can be weighted by time of day.

Time	TEMP °C		?		RAIN			WIND m/s			accumulated		
	Weight	Score	Weight	Result	Score	Weight	Result	Score	Weight	Result	Result	Weight	Result
0	3	2	3	100.0	1	1	1	1	1	1	105	15	1575
1	3	2	3	100.0	6	3	18	3	1	3	124	21	2604
2	3	7	3	14.3	3	2	6	5	1	5	28	18	509
3	3	8	3	12.5	4	2	8	7	2	14	38	21	788
4	3	3	3	100.0	5	3	15	10	2	20	138	24	3312
5	3	-1	3	100.0	6	3	18	12	3	36	157	27	4239
6	3	-1	3	100.0	7	3	21	15	3	45	169	27	4563
7	2	7	3	14.3	5	3	15	19	3	57	88	18	1589
8	2	8	3	12.5	4	2	8	23	3	69	92	16	1464
9	1	9	3	11.1	3	2	6	27	3	81	99	8	793
10	1	10	2	10.0	2	2	4	31	3	93	108	7	756
11	1	15	1	6.7	1	1	1	10	2	20	29	4	115
12	1	27	2	50.0	0	1	0	12	3	36	87	6	522
13	1	30	3	100.0	10	3	30	15	3	45	176	9	1584
14	1	27	2	50.0	0	1	0	19	3	57	108	6	648
15	1	16	1	6.3	0	1	0	23	3	69	76	5	381
16	1	10	2	10.0	0	1	0	15	3	45	56	6	336
17	1	12	2	8.3	0	1	0	12	3	36	45	6	272
18	2	9	3	11.1	0	1	0	10	2	20	33	12	397
19	2	8	3	12.5	0	1	0	9	2	18	33	12	390
20	2	5	3	100.0	0	1	0	5	1	5	107	10	1070
21	3	-3	3	100.0	0	1	0	4	1	4	107	15	1605
22	3	-5	3	100.0	0	1	0	3	1	3	106	15	1590
23	3	-10	3	100.0	0	1	0	2	1	2	105	15	1575
24	3	-3	3	100.0	0	1	0	1	1	1	104	15	1560

Fig. 15. Examples of a weather metric at different times of the day.

The weather metric, time, and categorised location result, when amalgamated with pre-disposition give a measure of risk. Put simply, if the monitored person is well rested and is outside on a sunny afternoon in a place which is defined as normal then perceived risk may be low, an accumulated score is used to derive the overall risk. When viewed together, these measures can be used to decide the point at which to override the privacy rule to ensure the monitored person is not harmed. The following section describes a working application designed to illustrate this.

6 Risk Analysis

Complex methods can be applied to calculate the perceived risk to the PwS; all the metrics described may be used to adjust the overall sensitivity of the system.

6.1 Inferring an Unknown Location

The Tensorflow Classifier, described in Sect. 3, is used in prediction of where the subject is in relation to normally visited places. If the subject moves to a new space, the contextual risk of that activity is assessed using time, distance from home, and forecasted weather conditions. In Fig. 16(a) movement along the test trajectory is outside known areas (shaded grey), distance and known temperature for the area is monitored (left graph above map). Risk is visualised in the right graph. As distance from home (start point) reduces, the system perceives this as returning and hence risk decreases. In Fig. 16(b) a detour outside a known path instigates appraisal and logs this as a new place, leading to an accumulation of risk, that is reset when the probability of normal movement increased, as shown in Fig. 16(c).

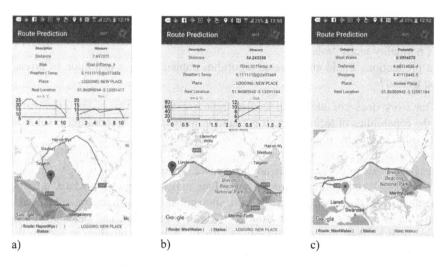

a)   b)   c)

Fig. 16. (a) Graphed representation of accumulating risk reducing with distance. (b) Risk increasing when taking a detour from the trained path. (c) Correct categorisation of trajectory with 99% accuracy – risk is reset.

The described system is indicative of how machine learning can be used to assess the 'normality' of outdoor movement and changes in sleep and heart rate patterns. The attraction of using AI in this way is that data learnt can be transferred across platforms and re-training can take place using a private network overnight. The tensor allows the mobile device carried by the PwS (in our study a phone) to act as an autonomous agent that does not require the internet, which has a very significant positive impact on battery life.

This research introduces a method using an AI agent to continuously assess the situation and make an ethical decision on overriding the default level of privacy. The point at which a measure of risk translates into a decision to breach privacy for the good of the monitored person is the topic of significant and ongoing debate, to which this technical study contributes. The following section briefly touches on considerations in this field.

7   Ethics

The ethical debate regarding the point at which location data is shared, and with whom is an interesting area to which our findings contribute. If activity, time, place, or weather is appraised as high risk or 'inappropriate', a prior moral framework that rates *safety and risk* versus *privacy* can justify that recent movement and current location may be shared. A wellbeing metric can be used to determine system sensitivity. The sharing of location can take the form of an SMS alert, or an alert via the internet including a map showing the current position of the PwS to a trusted carer. Continuous updates can facilitate speedy recovery. In all other cases, the PwS may continue independently and all data collected is kept private.

Several questions arise:

- When applied to vulnerable persons, who may decide the threshold and who defines what is 'inappropriate'?
- Is normality really 'safe'?
- In production, would an AI-based algorithm implementing a definition of privacy be trusted?

Our work does not attempt to answer these questions, but provides investigation into the capabilities of technology. It is found in literature that technological solutions fail to offer a considered approach to resolve well-known privacy issues. Surveillance of those who may be deemed vulnerable is considered by many as ethically inappropriate, but 'needs must' and carers are taking DIY approaches [38], with systems that use technology not optimised for privacy. This exposes them to potential security vulnerabilities as described above. We have shown that a fitness tracker can be used to learn what is normal in terms of heart rate while asleep. This, and other contextual matrices can be used to modify system sensitivity. A *private* monitoring system that uses AI to determine out-of-the-ordinary movement is novel. Since it respects privacy, this surveillance is *not intrusive*. Development and implementation of such a system is likely to provide PlwD with an ethically robust 'safety net' that may be used to improve quality of life. It can increase independent living of the PwS, provide peace of mind to the carer, while not requiring data sharing to call centres, or collection on central servers of tracking providers. Our system achieves secure data control and maintains data ownership.

8 Conclusion

The research presented shows promising results both in recognition of human geospatial activity and in prediction of movement along normally travelled routes. A wellness monitor discretely monitors both sleep and heart rate and this can be used to learn what is normal to enable it to flag exceptions. A cost-effective working prototype has been produced to demonstrate that deep-learning techniques can be applied to spatio-temporal data after programmatically categorising normally travelled trajectories. It has been found that when only part of a trajectory has been travelled, likely destinations can reliably be inferred. The application is designed to restrict personal information propagation to a home network and the limitations of computing capacity do not detract from the quality of results.

The World Health Organisation recognises that surveillance is intrusive, that the human rights of PwS are sometimes denied and that abuse of liberties is present. Locking doors to stop a person eloping violates their human right to liberty, but surveillance normally results in sharing of personal information, so is contrary to their human right to private life. Risk, when deviations from known places are sensed, is assessed automatically on a smart-phone in the context of time, extent and weather conditions.

Human rights (of private life and liberty) of the person with symptoms will be respected until the point at which it is judged that a prior moral argument of safety and risk supersedes the importance of privacy. If this happens, alerts containing location and recent movements are shared with an assigned carer, thus facilitating swift recovery.

The potential of the AI system described here is considerable. It is likely that many who value the importance of privacy highly will welcome a surveillance system that monitors but does not divulge detail. Predictions of likely trajectory of movement using real-time location data is novel, as is the concept of private surveillance as described. Availability of an internet connection or at least cellular coverage to deliver alerts is a requirement for implementation.

Ongoing work includes the processing of data from recruited volunteers, it is difficult to assess how the data-sets used differ from that which could be collected from PwS, subject to gaining the appropriate ethical approvals trials will embark with recruited PwS. The assessment of complex and intertwined trajectories and comparison of different scales of movement is currently under investigation. Findings will contribute to further refinement of the methodology after consultation with health professionals and PlwD. In an ideal scenario this would be used for prolonged independence of PwS, alleviation of a 24/7 burden of care, and could delay the necessity of moving the PwS to a care home.

Acknowledgements. Knowledge Economy Skills Scholarships (KESS) is a pan-Wales higher level skills initiative led by Bangor University on behalf of the HE sectors in Wales. It is part funded by the Welsh Government's European Social Fund (ESF) convergence programme for West Wales and the Valleys and is supported by the industrial partner SymlConnect Limited.

References

1. Williams, S., Müller, B.: Agents and dementia—smart risk assessment. In: Criado Pacheco, N., Carrascosa, C., Osman, N., Julián Inglada, V. (eds.) EUMAS/AT-2016. LNCS (LNAI), vol. 10207, pp. 277–284. Springer, Cham (2017). https://doi.org/10.1007/978-3-319-59294-7_22
2. Williams, S., Ware, J.M.: Is the use of 'mobile computer technology' appropriate for locating people with dementia? In: 2015 Proceedings of GIS Research UK (GISRUK), pp. 659–664. Figshare, Leeds (2015). https://doi.org/10.6084/m9.figshare.1491375
3. Read, S., Toye, C., Wynaden, D.: Experiences and expectations of living with dementia: a qualitative study. Collegian **24**(5), 427–432 (2017). https://doi.org/10.1016/j.colegn.2016.09.003
4. Silverstein, N.F.: Dementia and Wandering Behavior: Concern for the Lost Elder. Springer, New York (2006). ISBN 0-8261-0272-7
5. McShane, R., et al.: Getting lost in dementia: a longitudinal study of a behavioral symptom. Int. Psychogeriatr. **10**(03), 253–260 (1998). http://www.ncbi.nlm.nih.gov/pubmed/9785146
6. Ali, N., et al.: Risk assessment of wandering behavior in mild dementia. Int. J. Geriatr. Psychiatry **31**, 367–374 (2016). https://doi.org/10.1002/gps.4336
7. Martyr, A., et al.: Living well with dementia: a systematic review. Alzheimer's Dement. J. Alzheimer's Assoc. **13**(7), 1567–1568 (2017). https://doi.org/10.1016/j.jalz.2017.07.725
8. Cipriani, G., Lucetti, C., Nuti, A., Danti, S.: Wandering and dementia. Psychogeriatrics **14**, 135–142 (2014). https://doi.org/10.1111/psyg.12044
9. Riikonen, M., Mäkelä, K., Perälä, S.: Safety and monitoring technologies for the homes of people with dementia. Gerontechnology **9**(1), 32–45 (2010). https://doi.org/10.4017/gt.2010.09.01.003.00
10. Mathews, L.: Data From 540,000 GPS Vehicle Trackers Leaked Online forbes.com (2017). https://www.forbes.com/sites/leemathews/2017/09/22/data-from-540000-vehicle-tracking-devices-leaked-online/#40b9c009274b. Accessed 19 Sept 2018
11. Trackmageddon website: Multiple vulnerabilities in the online services of (GPS) location tracking devices (2018). https://0x0.li/trackmageddon/. 18 Aug 2018
12. World Health Organisation: Ensuring a human rights-based approach for people living with dementia (2015). The need for a human-rights based approach: http://www.who.int/mental_health/neurology/dementia/dementia_thematicbrief_human_rights.pdf. Accessed Mar 2018
13. LeCun, Y., Bengio, Y., Hinton, G.: Deep learning. Nature **521**, 436–444 (2015). https://doi.org/10.1038/nature14539
14. Schmidhuber, J.: Deep learning in neural networks: an overview. Neural Networks **61**, 85–117 (2014). https://doi.org/10.1016/j.neunet.2014.09.003
15. Brownlee, J.: How to Implement the Backpropagation Algorithm From Scratch in Python (2016). Machine Learning Mastery: https://machinelearningmastery.com/implement-backpropagation-algorithm-scratch-python/. Accessed Mar 2018
16. Hochreiter, S., Schmidhuber, J.: Long short-term memory. Neural Comput. **9**(8), 1735–1780 (1997). https://doi.org/10.1162/neco.1997.9.8.1735
17. Kwapisz, J.R., Weiss, G.M., Moore, S.A.: Activity recognition using cell phone accelerometers. ACM SIGKDD Explor. Newsl. **12**(2), 74–82 (2010). https://doi.org/10.1145/1964897.1964918

18. Ramamurthy, S.R., Roy, N.: Recent trends in machine learning for human activity recognition—a survey. Data Min. Knowl. Discov. **8**, 1–19 (2018). https://doi.org/10.1002/widm.1254

19. Arifoglu, D., Bouchachia, A.: Activity recognition and abnormal behaviour detection with recurrent neural networks. In: 14th International Conference on Mobile Systems and Pervasive Computing (MobiSPC 2017) (2017). Procedia Comput. Sci. **110**, 86–93. https://doi.org/10.1016/j.procs.2017.06.121

20. Batista, E., Borras, F., Casino, F., Solanas, A.: A study on the detection of wandering patterns in human trajectories. In: 6th International Conference on Information, Intelligence, Systems and Applications (IISA), pp. 1–6. IEEE, Corfu (2015)

21. Maneewongvatana, S., Mount, D.M.: Analysis of approximate nearest neighbor searching with clustered point sets. CoRR arXiv:cs/9901013v1 (1999)

22. SciPy.Org: Interpolation (scipy.interpolate). https://docs.scipy.org/doc/scipy/reference/interpolate.html. Accessed 17 Mar 2018

23. Google: Encoded Polyline Algorithm Format. Google Maps Api: https://developers.google.com/maps/documentation/utilities/polylinealgorithm?csw=1. Accessed 17 Mar 2018

24. Zheng, Y., et al.: Geolife GPS trajectories 1.1. In: Geolife GPS Trajectory Dataset - User Guide. Microsoft Research (2011). https://www.microsoft.com/en-us/research/publication/Geolife-gps-trajectory-dataset-user-guide. Accessed 18 Sept 2018

25. González, M.C., Hidalgo, C.A., Barabási, A.: Understanding individual human mobility patterns. Nature **453**, 779–782 (2008). https://doi.org/10.1038/nature06958

26. Google Maps Platform: Snap to Roads. Roads API: https://developers.google.com/maps/documentation/roads/snap. Accessed 18 Sept 2018

27. Fabrice Marchal: TrackMatching API. TrackMatching Website: https://mapmatching.3scale.net/mmswag. Accessed 18 Sept 2018

28. Google Directions API: Google Directions API. Google Maps: https://developers.google.com/maps/documentation/directions/start. Accessed 18 Sept 2018

29. Lambertus: YOURS Routing_API. Yet another OpenStreetMap Route Service: https://wiki.openstreetmap.org/wiki/YOURS#Routing_API. Accessed 18 Sept 2018

30. Laure, P.-D., Yammine, P., Bastuji, H., Croisilef, B.: Sleep and Alzheimer's disease. Sleep Med. Rev. **19**, 29–38 (2015)

31. Hope, T., et al.: Predictors of institutionalization for people with dementia living at home with a carer. Int. J. Geriatr. Psychiatry **13**(10), 682–690 (1998)

32. McCurry, S., Logsdon, R., Teri, L., Vitiello, M.: Sleep disturbances in caregivers of persons with dementia: contributing factors and treatment implications. Sleep Med. Rev. **11**(2), 143–153 (2007)

33. Brodaty, H.D.: Family caregivers of people with dementia. Dialogues Clin. Neurosci. **11**(2), 217–228 (2009). https://www.ncbi.nlm.nih.gov/pmc/articles/PMC3181916/. Accessed 17 Apr 2018

34. Keill, A.K., et al.: Validity of wearable fitness trackers on sleep measure. Med. Sci. Sports Exerc. **48**(5S), 10 (2016)

35. scikit-learn developers: sklearn.cluster.AgglomerativeClustering. http://scikit-learn.org/stable/modules/generated/sklearn.cluster.AgglomerativeClustering.html. Accessed 18 Sept 2018

36. sklearn.cluster developers: sklearn.cluster.KMeans. scikit-learn: http://scikit-learn.org/stable/modules/generated/sklearn.cluster.KMeans.html. Accessed 18 Sept 2018

37. Dark Sky Website: Dark Sky Weather API. https://darksky.net/dev. Accessed 18 Sept 2018

38. Gibson, G., et al.: The everyday use of assistive technology by people with dementia and their family carers: a qualitative study. BMC Geriatr. **15**, 89 (2015). https://doi.org/10.1186/s12877-015-0091-3

Active Learning for Conversational Interfaces in Healthcare Applications

Aki Härmä[1]([⊠]) [iD], Andrey Polyakov[2] [iD], and Ekaterina Artemova[3] [iD]

[1] Philips Research, Eindhoven, The Netherlands
aki.harma@philips.com
[2] Philips Research, Moscow, Russia
[3] Sberbank, Moscow, Russia

Abstract. In automated health services based on text and voice interfaces, there is a need to be able to understand what the user is talking about, and what is the attitude of the user towards a subject. Typical machine learning methods for text analysis require a lot of annotated data for the training. This is often a problem in addressing specific and possibly very personal health care needs. In this paper, we propose an active learning algorithm for the training of a text classifier for a conversational therapy application in the area of health behavior change. A new active learning algorithm, Query by Embedded Committee (QBEC), is proposed in the paper. The methods are particularly suitable for the text classification task in a dynamic environment and give a good performance with realistic test data.

1 Introduction

The application context of the current paper is the development of automated therapeutic conversational interventions for behavior change [2], in particular, related to substance abuse. Counseling is known to be the most effective intervention for many lifestyle diseases, but counseling sessions are expensive for the health care system and often inconvenient for patients. Automation of the effective mechanisms of counseling by automated agents would lead to better coverage and cost savings. In a typical application, a conversational agent would implement some elements of the Cognitive Behavioral Therapy [9]. Typically, the agent would be available through a social media platform possibly with a speech interface. The text understanding system should be able to detect the topics and sentiment structures relevant to the control of the conversation according to the selected therapeutic strategy.

Recurrent neural networks are popular for text understanding, but they require a large corpus of labeled training data, which is difficult to collect. Also, natural language communication is an example of a non-stationary learning environment where the evolution in the conversational culture over time and populations require local customization and maintenance of the classifier, possibly even at the level of an individual customer. The client talk related to a particular substance may be very specific and patients may even develop a personal

F. Koch et al. (Eds.): AIH 2018, LNAI 11326, pp. 48–58, 2019.
https://doi.org/10.1007/978-3-030-12738-1_4

vocabulary to discuss about the addiction. Also, the content is naturally very sensitive and it is therefore desired that the client talk should not be uploaded to a cloud processing platform but processed locally in the client device.

One approach for the maintenance and continuous improvement of a classifier in the production environment is to use *active learning* (AL) methods [3,16]. In pool-based AL methods, only a small part of the available content is manually labeled and used to train the classifier. A typical approach is to use a committee of classifiers [12] to select items that are *difficult to classify* based on the current statistics. This approach works well in many conventional problems but often leads to robustness problems that are common in many deep learning architectures [6]. Also, while the classification of client talk may be, at least in the future Edge AI technologies, performed the client device, the detection of novel training content selection based in AL in a client device is significantly more challenging.

In this paper, we demonstrate an application of active learning in the classification of short text messages, *tweets*, from a social media platform using a text classifier based on Recurrent Neural Networks, RNNs [8]. We propose an algorithm for the pool-based selection where the committee method is applied in a latent variable space. In particular, the committee is embedded in a space spanned by the class likelihoods of the last classifier. In this paper, the method is called Query by Embedded Committee, QBEC. The method is computationally significantly lighter than the conventional Query-by-Committee, QBC, method.

2   Sample Selection Methods

In pool-based active learning [16] new samples are selected to the training data from a large pool of unlabeled content. The selection may be based on different principles and aim at selecting the most informative or representative samples [10,17], reduce the variance of the classification errors [3], or *diameter* in a space spanned by alternative classifiers [4].

The AL process starts with an initial set P_0 of labeled tuples of K feature vectors x_k and corresponding labels l_k, i.e.,

$$P_0 = \{x_k, l_k\}, k = 0, \cdots, K - 1 \tag{1}$$

The Initial classification model M_0 is developed using P_0. Next, a new set S_1 is selected from the pool. The samples are manually labeled by a human oracle, for example, a health counselor. The new training data P_1 is produced by adding the samples S_1 to P_0. The model is updated and deployed. The same update cycle can then be continuously repeated.

The selection of the next batch S_{j+1} of B samples can be based on many different criteria. The minimum requirements for a jth iteration are

1. *novelty:* $P_j \cap S_{j+1} = \emptyset$
2. *richness:* $x_n \neq x_m, \forall n, m \in S_{j+1}$

i.e., the B new samples in S_{j+1} should be novel and they should be different from each other.

2.1   Query by Committee

In the popular Query-by-Committee (QBC) method [12] the novelty condition is addressed by measuring the disagreement in a committee of R different classifiers C_r trained using P_j.

$$d_j = \mathcal{D}\left[C_0(x_k), C_1(x_k), \cdots, C_R(x_k)\right] \tag{2}$$

where $\mathcal{D}[]$ is some measure to compute the disagreement.

In a typical case, the disagreement is based on vectors of class likelihoods given by the classifiers $\mathbf{p}_r(x_k) = C_r(x_k)$. In a committee of two classifiers, the disagreement can be defined as a norm of the difference $d_k = |\mathbf{p}_0(x_k) - \mathbf{p}_1(x_k)|$.

Algorithm 1. Query by Committee

Require: Label the initial data set P_j and set $j = 0$
1: **repeat**
2:    Train the main classifier model M_j using P_j data
3:    Train the committee classifiers C_0, C_1 using P_j data
4:    Compute the disagreement $d_k \in S$ in the new batch
5:    Pick K samples from S which has the highest disagreement using the knockout neighborhood penalization described above.
6:    Labeling of the new samples by a human expert
7:    Add new samples to the training data $P_{j+1}, j = j + 1$
8: **until** Stopping criteria are met.

The committee often disagrees on very similar samples, and therefore the basic algorithm does not provide the required *richness* for the new sample collection. A *pareto* optimal solution is needed to meet both the *novelty* and *richness* conditions. The richness is related to the nearest neighbor problem (NN). The k-nearest-neighbor searching problem (kNN) is to find the k -nearest points in a dataset $X \subset \mathbb{R}^d$ containing n points to a query point $q \in \mathbb{R}^d$ under some norm. There are several effective methods for this problem when the dimension d is small (e.g. 1, 2, 3), such as Voronoi diagrams [18] or Delaunay triangulation [5]. When the dimension is moderate (e.g., up to the 10's), it is possible to use kd trees [8] and metric trees [11]. If the dimension is high, then Locality-Sensitive Hashing (LSH) is the very popular method used in applications. In the current paper, we use an iterative algorithm where the new samples that are close to already selected samples are penalized. Experiments with other sampling principles is a part of future work.

2.2   Query by Embedded Committee

The selection of the new samples based on a disagreement of a committee assumes a certain variability among the committee members [12]. This is typically achieved by using different initialization of the classifiers C_j, or by using different classifier prototypes or kernels. In the case of a complex model, for example, based on multiple layers of memory networks and dense layers, the training of a committee can be a large effort and may take, for example, several hours of processing time in a GPU. In principle, the training of each committee member model takes as much time as the training of the main model itself. However, the final scoring of the network is light and can be performed, for example, in a smartphone or another end-user device.

In this paper, the proposed method is to use the committee in another feature space derived from the outputs of the model. A multi-class classifier is often developed using the one-hot encoding principle where the classifier produces a vector of class likelihoods $\mathbf{p}_r(x_k) = M(x_k)$ for a feature vector x_k. The likelihoods represent the class predictions for the testing data. In a geometric sense, the likelihood vectors $\mathbf{p}_r(x_k)$ span an orthonormal space, a *class space* of the current classifier, where each axis represents a class.

The proposed method is a variation of the QBC method where the selection task is performed in the class space of the current classifier. The class space is a metric low-dimensional space, and there the committee can be based on conventional classification tools, e.g., based on a random forest or another relatively light algorithm. The training of the committee of classifiers and testing of them on a new data can be performed in an end-user device. Therefore, this enables local active learning of the classification model.

The processing steps of the proposed method are described in Algorithm 2 below.

Algorithm 2. Query by Embedded Committee (QBCSC)

1: **repeat**
2:    Use classifier M_j to get class likelihood vectors for data $\mathbf{p}_r(x_k)\forall x_k \in P_j$
3:    Use QBC method defined in the class space to select the new samples for labeling.
4:    Add new samples to the training data $P_{j+1}, j = j + 1$
5: **until** Stopping criteria are met.

In this paper we call the modified method Query by Embedded Committee (QBEC), to separate it from the conventional QBC. In the current paper, the committee is *embedded* in the class space. Naturally the same can also be performed in another output space, for example, corresponding to intermediate layers of the network.

2.3   Computational Load

QBEC method works faster then QBC due to hypothesis space reduction. Namely, it has been shown in [7] that the number of queries for labels that the algorithm will make is $\mathcal{O}(\frac{d}{g}log(\frac{1}{\varepsilon}))$, where d is the Vapnik-Chevonenkis dimension, g is some constant, ε is required accuracy.

If QBC works in $\mathbb{R}^n$, then $d = n + 1$ (as the hypothesis space is divided by set of oriented hyperplanes). Simultaneously, QBEC works in $\mathbb{R}^k$, where $k \ll n$

is the dimension of a space which is spanned by the class likelihoods of the current classifier. So, the corresponding Vapnik-Chervonenkis dimension will be $m = k + 1$. So, the complexity of QBEC is $\mathcal{O}(\frac{m}{g}log(\frac{1}{\varepsilon})) < \mathcal{O}(\frac{d}{g}log(\frac{1}{\varepsilon}))$.

3   Experiments

Let us start with a synthetic example to illustrate the differences between QBC and the proposed QBEC method, and their benefits over random sampling from the pool.

3.1   Synthetic Example

The original synthetic data is shown in Fig. 1(a) with two classes illustrated by red crosses and blue circles. A random forest (RF) classifier was designed for the set P_0 with 100 labeled samples. In the QBC method a committee of two RF classifiers was designed using a different initialization. The selection of new samples was based on selection of the samples with the largest difference in the class likelihood values between the classifiers. Figure 1(b) shows an example of a

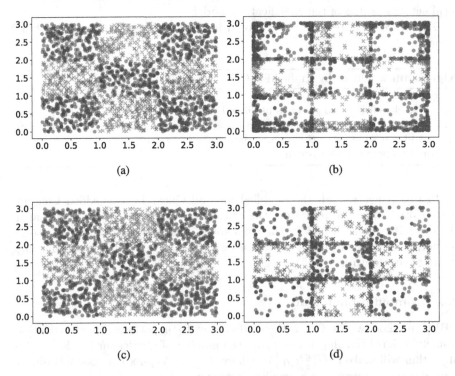

Fig. 1. (a) test data, (b) QBC samples, (c) random samples, (d) QBEC sampling. The x and y axes are arbitrary coordinates. (Color figure online)

QBC sampling. A random sampling used in the reference condition is illustrated in Fig. 1(c). The QBC method clearly takes more samples from the class borders that the random sampling method. The QBEC method also focuses on the class borders but puts more emphasis on the borders between classes rather than outer borders.

The accuracy in the training in the three methods is shown in Figs. 2(a) and 2(b). The QBC and QBEC have a similar performance in the first batches, but the accuracy of QBEC method keeps improving at the point where the performance of QBC saturates. This may be understood in this case by comparing the selections in the two methods in Fig. 1(b). In the QBCSC the sampling focuses on borders between the classes while in the conventional QBC solution a large number of samples are selected from the outskirts of the feature space which is less relevant for the class confusions measured by the accuracy.

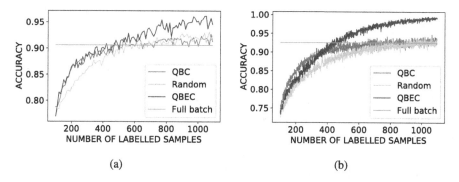

(a)   (b)

Fig. 2. Accuracy in an iterative active learning experiment using the three methods. The straight green line corresponds to the accuracy of a single classifier trained with the full set of 600 labeled samples. Batch sizes are (a) 1 (b) 10 (Color figure online)

4   Experiment with Tweets

In this paper, the content is from *Twitter*, which is a popular short-text messaging platform. The content was selected by keywords that relate to smoking and tobacco use. In the typical flow of content, the test system gave approximately 1000 tweets per day when excluding repetitions (*re-tweets*) of the same message (Fig. 3).

In the current paper the content is manually classified into three classes: *sustain talk, change talk*, and *neutral* communication. The two first classes are considered important elements in many therapeutic techniques for substance abuse, such as CBT [9] or Motivational Interviewing (MI) [14]. The target behavior is to reduce or quit smoking. Sustain and change talk contains all client talk that speaks against or for the target behavior, respectively. The neutral class contains all other content with the same keywords. The data contains all English language

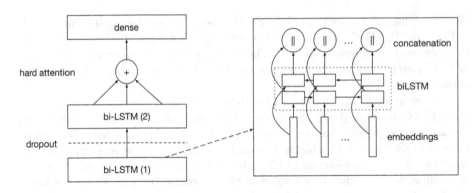

Fig. 3. Architecture of the deep learning neural network used for text classification. **On the left**: the general architecture of the deep learning neural network. **On the right**: the input layer and the first Bi-LSTM Layer. The second Bi-LSTM copies the first one, except the input are not embeddings, but the output of the first Bi-LSTM layer. See Table 1 for the number of units and other parameters.

messages from Oct. 2017 until the end of Jan 2018[1]. There are cultural elements in the data. For example, the tweets from October contain messages that relate to the *Stoptober* smoking cessation campaign in the UK and other countries, in December there are tweets from people who plan to *quit for January*, and there are several referrals to a popular song called *cigarette daydreams*.

4.1 Text Classification System

In this paper, we use a typical architecture for a text classifier based on a state-of-the-art deep learning RNN tools. The text classifier model has six components, presented in Table 1.

Table 1. Architecture of the deep learning neural network used for text classification

Layer	Parameters
Input: embedding layer	Google SGNS [13], Stanford GloVe [15]
Bi-LSTM	64 LSTM units
Dropout	0.2
Bi-LSTM	32 LSTM units
Hard attention layer [1]	
Output: dense layer	Regular dense layer with softmax activation

---

[1] The ethical and legal approval of the data collection was granted, and handled according to, by the Internal Committee for Biomedical Experiments (ICBE) of Philips.

The embedding layer is meant to map each word of the input text into a low dimensional embedding vector, while the bidirectional layers get higher level features from the input, dropout being used for regularization. The hard attention layer is used for global re-weighting of hidden layers, and the desired class label is chosen using a regular dense layer with softmax activation.

5   Results

The initial classifier $C^{(0)}$ was trained using a manually classified set of 2398 tweets. Examples of typical tweet types and their counts in the initial training set are shown in Table 2. Also, an independent test data set with manually labeled tweets was used for testing. The performance of $C^{(0)}$ in an independent training set is poor; the accuracy is barely above 0.5. In the following experiment, the active learning process was executed sequentially so that the current dump of tweets about the target topic was downloaded once a day, classified using classifier $C^{(n)}$. Approximately one percent, typically around 30 tweets, were selected to the manual labeling using one of the selection methods. The samples were manually labeled and included in the training set, and subsequently used in the training of the next model $C^{(n+1)}$.

Table 2. Examples of tweets and their counts in the initial training set.

Talk type	Example	#samples
Change talk	Two weeks without smoking!	246
Sustain talk	I'm having a cigarette	514
Neutral	A man was smoking outside	1651

The latent space representation formed by the outputs of likelihoods at the output layer of the network is illustrated in Fig. 4. The three talk types are separated in the latent variable space.

The numbers of new labeled samples resulting from daily 23 iterations in the three methods are illustrated in Figs. 5a–c. First, it seems that random selection rarely picks samples from change talk category while those are much more common in the two other methods. The accuracy in the three methods, respectively, is shown in Fig. 5d. In the random selection the accuracy does not improve over the iterations, but in the two other methods, there is a clear improving trend. Unlike the results with the checker board data, there is not really a difference in accuracy between QBC and QBEC, although, the computational requirements and processing time in QBEC is obviously significantly lower than in QBC.

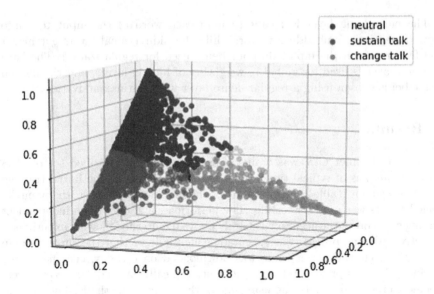

Fig. 4. Example of three classes of tweets in the 3D latent variable space.

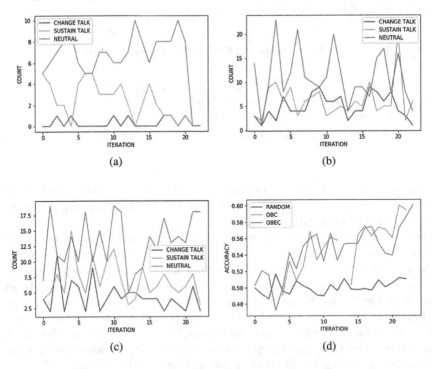

Fig. 5. (a) Random selection (b) QBC, and (c) QBEC, (d) accuracy in the three methods based on correct detections. X-axis represents the iteration number (days of downloaded tweets). One data point in QBC curve is missing due to an error in data handling.

6 Conclusions

In this paper, we propose a new algorithm for active learning in the application of text classification in the application of health counseling. The training of a classifier for a specific complex talk type requires a large labeled database which is typically difficult, expensive, and time-consuming. There may also continuous concept drift in the target area, for example, due to various cultural influences.

A popular approach for active learning is to use a disagreement in a committee of classifiers to select samples for manual labeling and inclusion into the training. These methods are commonly called Query-by-Committee (QBC) methods. The QBC methods require that multiple classifiers are trained for the task. In applications where the classifier is complex, a.e.g, a deep neural network model, and requires a long training time this may be problematic. In the method introduced in the current paper, the committee selection is performed in a low dimensional space spanned by the likelihoods of the current classifier model. In this case, the actual classifiers of the committee can be fairly simple. The method is called Query-by-Embedded-Committee (QBEC).

We demonstrate the performance of QBEC first using synthetic data. The performance of QBEC turns out to be superior to the random selection of training samples and it, surprisingly, exceeds the performance of QBC. One may speculate that this is because the embedding based on the prediction likelihoods inherently zooms the committee to *zoom* into areas where the disagreement is largest.

In a second experiment, we trained a complex classifier for classification of tweets related to smoking behavior into three classes. The classes represent change talk, sustain talk, and neutral communication of the talker about tobacco use. This is a very challenging classification problem requiring a large labeled database. In the active learning experiment, 1% of tweet content downloaded on each day was manually labeled and included in the new model. It was shown that QBC and QBEC outperform random selection of samples. However, the results of the two methods are similar. However, it should be noted that the computational of QBEC is significantly lower than in QBC. Therefore, the sample selection in QBEC could be performed even in a customer device such as a smartphone.

References

1. Bahdanau, D., Cho, K., Bengio, Y.: Neural machine translation by jointly learning to align and translate. arXiv preprint arXiv:1409.0473 (2014)
2. Bickmore, T., Giorgino, T.: Health dialog systems for patients and consumers. J. Biomed. Inform. **39**(5), 556–571 (2006). https://doi.org/10.1016/j.jbi.2005.12.004
3. Cohn, D.A., Ghahramani, Z., Jordan, M.I.: Active learning with statistical models. J. Artif. Int. Res. **4**(1), 129–145 (1996). http://dl.acm.org/citation.cfm?id=1622737.1622744
4. Dasgupta, S.: Coarse sample complexity bounds for active learning. In: NIPS 2005, pp. 235–242. MIT Press, Cambridge (2005). http://dl.acm.org/citation.cfm?id=2976248.2976278

5. Delaunay, B.: Sur la sphere vide. a la memoire de georges voronoi. Bulletin del'Academie des Sciences del'URSS (6), 793–800 (1934)
6. Fawzi, A., Moosavi-Dezfooli, S.M., Frossard, P.: The robustness of deep networks: a geometrical perspective. IEEE Sig. Process. Mag. **34**(6), 50–62 (2017). https://doi.org/10.1109/MSP.2017.2740965
7. Freund, Y., Seung, H.S., Shamir, E., Tishby, N.: Selective sampling using the query by committee algorithm. Mach. Learn. **12**(28), 133–168 (1997)
8. Hochreiter, S., Schmidhuber, J.: Long short-term memory. Neural Comput. **9**(8), 1735–1780 (1997)
9. Hofmann, S.G., Asnaani, A., Vonk, I.J., Sawyer, A.T., Fang, A.: The efficacy of cognitive behavioral therapy: a review of meta-analyses. Cogn. Ther. Res. **36**(5), 427–440 (2012). https://doi.org/10.1007/s10608-012-9476-1, https://www.ncbi.nlm.nih.gov/pmc/articles/PMC3584580/
10. Hoi, S.C.H., Jin, R., Lyu, M.R.: Batch mode active learning with applications to text categorization and image retrieval. IEEE Trans. Knowl. Data Eng. **21**(9), 1233–1248 (2009). https://doi.org/10.1109/TKDE.2009.60
11. Liu, T., Moore, A.W., Gray, A., Yang, K.: An investigation of practical approximate nearest neighbor algorithms. In: Saul, L.K., Weiss, Y., Bottou, L. (eds.) Advances in Neural Information Processing Systems, pp. 825–832. MIT Press, Cambridge (2015)
12. McCallum, A., Nigam, K.: Employing EM and pool-based active learning for text classification. In: ICML 1998, pp. 350–358. Morgan Kaufmann Publishers Inc., San Francisco (1998). http://dl.acm.org/citation.cfm?id=645527.757765
13. Mikolov, T., Chen, K., Corrado, G., Dean, J.: Efficient estimation of word representations in vector space. arXiv preprint arXiv:1301.3781 (2013)
14. Miller, W.R., Rose, G.S.: Toward a theory of motivational interviewing. Am. Psychol. **64**(6), 527–537 (2009). https://doi.org/10.1037/a0016830, http://www.ncbi.nlm.nih.gov/pmc/articles/PMC2759607/
15. Pennington, J., Socher, R., Manning, C.: Glove: global vectors for word representation. In: Proceedings of the 2014 Conference on Empirical Methods in Natural Language Processing (EMNLP), pp. 1532–1543 (2014)
16. Tong, S., Koller, D.: Support vector machine active learning with applications to text classification. J. Mach. Learn. Res. **2**, 45–66 (2002). https://doi.org/10.1162/153244302760185243
17. Wang, L., Hu, X., Yuan, B., Lu, J.: Active learning via query synthesis and nearest neighbour search. Neurocomputing **147**(Suppl. C), 426–434 (2015). https://doi.org/10.1016/j.neucom.2014.06.042, http://www.sciencedirect.com/science/article/pii/S0925231214008145
18. Ying, S., Xu, G., Li, C., Mao, Z.: Point cluster analysis using a 3D Voronoi diagram with applications in point cloud segmentation. Int. J. Geo-Inf. (ISPRS) **4**(3), 1480–1499 (2015)

Analysis of Topic Propagation in Therapy Sessions Using Partially Labeled Latent Dirichlet Allocation

Ilyas Chaoua[1] , Sergio Consoli[2](✉) , Aki Härmä[2] , Rim Helaoui[2] ,
and Diego Reforgiato Recupero[1]

[1] Mathematics and Computer Science Department, University of Cagliari,
Via Ospedale 72, 09124 Cagliari, Italy
ilyaschaoua@gmail.com, diego.reforgiato@unica.it
[2] Philips Research, High Tech Campus 34, 5656 AE Eindhoven, The Netherlands
{sergio.consoli,aki.harma,rim.helaoui}@philips.com

Abstract. The full comprehension of how topics change within psychotherapeutic conversation is key for assessment and therapeutic strategies to adopt by the counselor to the patients. That might enable artificial intelligence (AI) approaches to recommend the most suitable strategy for a new patient. Basically, understanding the topics dynamics of previous cases allows choosing the best therapy to perform for new patients depending on their current conversations.

In this paper we leverage Partially Labeled Dirichlet Allocation with the goal to detect and track topics in real-life psychotherapeutic conversations. On the one hand, the detection of topics allows us identifying the semantic themes of the current therapeutic conversation and predicting topics ad-hoc for each talk-turn between the patient and the counselor. On the other hand, the tracking of topics is key to understand and explore the dynamics of the conversation giving insights and tips on logic and strategy to adopt.

We point out that the entire conversation is structured and modeled according to a sequence of ongoing topics that might propagate through each talk-turn. We present a new method that combines topic modeling and transitions matrices that gives important information to counselors for their therapeutic strategies.

Keywords: Conversational AI · Psychotherapeutic conversations ·
Topics detection and modeling ·
Partially Labeled Dirichlet Allocation · Transitions matrices

1 Introduction

Analysis and research of therapeutic conversations is a growing domain of research: technological advances and innovations in areas such as Natural Lan-

---

Authors are listed in alphabetic order since their contributions have been equally distributed.

© Springer Nature Switzerland AG 2019
F. Koch et al. (Eds.): AIH 2018, LNAI 11326, pp. 59–75, 2019.
https://doi.org/10.1007/978-3-030-12738-1_5

guage Processing (NLP), Semantic Web, Big Data, Artificial Intelligence and healthcare have triggered such a growth [13].

Humans interactions have been targeted and machine learning approaches have been employed to detect unusual patterns. An example, authors in [6] aimed to infer predictive models to structure task-oriented dialogs.

Cognitive Behavior Therapy (CBT) entails therapeutic conversational methods which consist of therapies that aim at treating mental health problems, emotional challenges, sleeping difficulties, relationship problems, drug and alcohol abuse, anxiety and depression. Such therapies tackle and try to change the way of thinking and behaving of patients. Figure 1 shows a diagram of the CBT rationale[1].

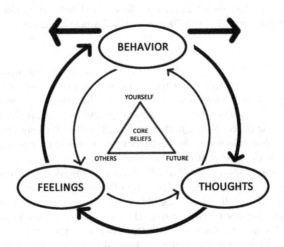

Fig. 1. The diagram depicts how emotions, thoughts and behaviors are related to each other. The inner triangle represents CBT's tenet that all humans' core beliefs can be summed up in the three mentioned categories.

More in detail, these therapeutic methods work by changing people's attitudes and their behavior by focusing on the thoughts, images, beliefs and attitudes that are held (a person's cognitive processes) and how these processes relate to the way a person behaves, as a way of dealing with emotional problems. The treatment is relatively short, taking five to ten months for most emotional problems. Patients usually attend one session per week where each session lasts less than 1 h. It is usually a face-to-face interaction between the counselor and the patient where the former needs to understand patient's feelings, e.g. confident, anxious, or depressed, as well as the causes of his feelings.

The conversation consists of a series of spoken sentences. Each is characterized by a certain topic: this creates a thematic structure to the whole therapeutic

---

[1] Image taken from Wikipedia https://en.wikipedia.org/wiki/Cognitive_behavioral_therapy.

conversation. The counselor employs techniques and strategies coming from clinical practice: he/she reacts to the patient and drives the conversation towards certain themes in order to tackle and solve the psychological problems of the patient.

It follows that the analysis and modeling of these human-to-human dialogues may be useful for the development of AI-based dialogue systems able to recommend the most appropriate therapeutic strategy to adopt by the practitioner for a new patient [6]. In such a context, a subset of the NLP research is related to the problem of topic detection and tracking (TDT), which has been widely focused and studied and combined with AI methods in the literature [15]. One goal of the TDT is the identification of the new topics in a conversation and their reappearance. Authors of [12] provide extensive background about that.

In this paper we focus on the conversation between counselor and patient and model the propagation of the topics identified during a given therapeutic conversation by using Partially Labeled Latent Dirichlet Allocation (PLDA) [21].

Traditional Latent Dirichlet Allocation (LDA) is one of the most popular topic model in the literature. LDA is based on a bag-of-words approach and is a generative statistical model that allows sets of observations to be explained by unobserved groups that explain why some parts of the data are similar. Our choice reflected on PLDA because the data we have used in our study are partially labeled and PLDA tends to achieve higher precision than traditional LDA on them.

The dataset we have used consists of 1729 real-life transcribed psychotherapeutic conversations, each made of different talk-turns. Further details about the data will be given in Sect. 4. Our approach works as it follows:

- First, we identify the most common topics used within the dataset.
- Then, the PLDA model takes as input the given conversations and detects significant words for each topic.
- The trained PLDA model is thus able to determine the potential topic addressed in each talk-turn.
- Within each conversation, the talk-turns flow is then transformed into a sequence of potential topics.
- Finally, for each topic, the semi-supervised PLDA topic model is evaluated by computing its coherence over the most significant words.

Our ultimate goal is to detect the key patterns within therapeutic conversations and to identify the topic switches according to the adopted dialogue strategy and topics propagation dynamics. In our method we are able to distinguish the topic changes driven by the counselor and the ones prompted by the patient. Two topic transition matrices are constructed accordingly to evaluate the two different topic changes. These matrices provide a numerical summary of the conversation and can be exploited to obtain tips for the overall understanding of the topics propagation dynamics.

This paper is further structured as it follows. Related works and literature background on automatic topic detection methods and therapeutic dialogue analysis are discussed in Sect. 2. LDA-based topic modelling algorithms are presented in Sect. 3, where we describe how they work, specifying their main characteristics and giving details about them. Section 4 includes details of the dataset we have employed for the experiments and the preprocessing we have performed on that. Our approach on TDT is presented in Sect. 5. Section 6 includes details related to the evaluation of our approach. Finally Sect. 7 ends the paper with conclusions and directions where we are headed.

2   Related Work

There is earlier work in the area of computational analysis of therapeutic sessions using topic modeling techniques, see, e.g., [5, 17]. However, there seems to be few earlier works on dynamics of topic propagation in those conversations.

Digitalizing spoken interactions and recommending specific treatments are two current trends within the therapeutic conversations research. They use effective NLP technologies with the goal of extracting knowledge in text form from consultation transcripts. For example, work in [3] discussed an idea to combine communication theory used in healthcare and a visualization text analytic technique called *Discursis*, with the goal of analyzing the conversational behavior in consultations. More specifically, Discursis[2] is a computer-based tool for analysing human communication that can assist practitioners in understanding the structure, information content, and inter-speaker relationships that are present within input data. Discursis processes conversation transcript data to determine the conceptual content of each conversation turn. It offers visualizations and reports on the above information such as a concept map of the communication content, communication channels, concept recurrence matrix, score cards for each conversation in terms of, e.g., leader, follower, innovator, promoter.

The classification of conversations is not an easy task in medical consultations as it includes intense performance requirements, it is time-consuming and it suffers from non-standardized annotating systems. Authors of [14] presented an automated annotating system which leverages a Labeled LDA model [20] to assess the relationships between a certain conversation and its annotations. Annotations are related to the subjects symptoms present within the therapeutic conversations. The system is therefore able to identify the relevant annotations in separate talk-turns.

LDA also been used in a different way by authors in [16]. In particular, they have analysed a LDA topic model [8] as an automatic annotator tool for the topics and therapy prediction of the conversation. One assumption made by the authors was related to the fact that the automated detection of topics can be used to predict factors such as patient satisfaction and ratings of the therapy quality rather than predict the symptoms. The employment of the Labeled LDA and the LDA indicates that the identification and tracking of topics can provide important information to clinicians. They can use such information to better assist the patients and improve their treatments.

---

[2] http://www.discursis.com/.

Other authors analysed human communications in [2] where they developed a discourse visualization system which converted transcribed conversations to time series, a text analysis model and a set of quantitative metrics to detect and assess significant features. Their system was able to identify the topics adopted in a certain discussion by the participants and generate reports for each conversation.

These metrics can be seen as an extension of recurrence quantification analysis into the symbolic domain. The proposed technique may be used to monitor the state of a communication system and inform about interaction dynamics, including the level of topic consistency between participants; the timing of state changes for the participants as a result of changes in topic focus; and, patterns of topic proposal, reflection, and repetition.

Other researchers proposed one more use of the LDA model. In particular they adopted a conceptual dynamic latent Dirichlet allocation (CDLDA) model for TDT in conversational text content [24]. The differences between traditional LDA and the CDLDA model is that whereas the former employs bag-of-words techniques to identify topics, the latter considers information such as speech acts, semantic concepts, and hypernym definitions in E-HowNet [11][3]. The proposed method extracts the dependencies between speech acts and topics, where hypernym information makes the topic structure more complete and extends the abundance of original words. Results performed by the researchers proposing this idea indicated that the approach outperforms the conventional Dynamic Topic Models [7], LDA, and support vector machine models, achieving very high performance for TDT.

Work performed in [1] includes OntoLDA for the task of topic labeling. OntoLDA adopts an ontology-based topic model and a graph-based topic labeling method. Basically, the topic labeling method is based on the ontological meaning of the concepts included in the discovered topics. This approach indicated each topic as a multinomial distribution of concepts, and each concept as a distribution of words. OntoLDA scaled better the topics coherence score than the classical LDA. This was achieved by combining ontological concepts with probabilistic topic models towards a combined framework applied to various types of text collections.

One more approach that improved the human-agent dialogs was presented by authors in [9]. Their approach leveraged the basis of contextual knowledge provided by Wikipedia category system. To build their approach they had to map the different utterances to Wikipedia articles and define their relevant Wikipedia categories as a list of topics. It followed that the detection method was able to recognize a topic without holding a priori knowledge of its subject category.

Authors in [16] questioned the use of LDA to cast more light on the role of topic modeling to provide a measure of content more general than word features with the goal to identify patient satisfaction and evaluations of therapy quality. The unsupervised model they introduced produces models similar to manual annotation, and it appears to be better at predicting evaluations of the therapeutic relationship and important features of communication style particularly

---

[3] http://ckip.iis.sinica.edu.tw/taxonomy.

that of the counselors. This may suggest that unsupervised models used in this way are able to discover and track topics to provide more insight to therapists, enabling them to better direct their conversations in time-limited consultations, and serve the identification of patients who can afterward be at risk of relapse or non-adherence to treatment.

3 Topic Modeling

Topic models are a family of probabilistic approaches that aim at discovering latent semantic structures in large documents. Based on the presumption that meanings are relational, they interpret topics or themes within a set of documents originally constructed from a probability distribution over words. As a result, a document is viewed as a combination of topics, while a topic is viewed as a blend of words.

One of the most widely used statistical language modeling for this end is Latent Dirichlet Allocation (LDA) introduced by Blei et al. [8]. LDA is a generative approach. It assumes that documents in a given corpus are generated by repeatedly picking up a topic, then a word from that topic according to the distribution of all observed words in the corpus given that topic. LDA aims at learning these distributions and inferring the (hidden) topics given the (observed) words of the documents [18]. Given the nature of our data which includes partial annotations, we employ the following two variants of LDA.

Labeled Latent Dirichlet Allocation (LLDA) [20] is a supervised version of LDA that constraints it by defining a one-to-one correspondence between topics and human-provided labels. This approach, illustrated in the probabilistic graphical model in Fig. 2, allows LLDA to learn word-label correspondences.

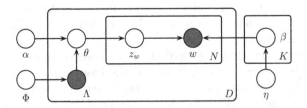

Fig. 2. Probabilistic graphical model of LLDA: unlike standard LDA, both the label set Λ as well as the topic prior α influence the topic mixture θ. β represents a vector of the parameters of the multinomial distribution whereas η are the parameters of the word prior. ω is the word, z is the per-word label assignments and ϕ is the label prior. Please check [20] for further details.

Partially Labeled Latent Dirichlet Allocation (PLDA) [21] is a semi-supervised version of LDA which extends it with constraints that align some learned topics with a human-provided label. The model exploits the unsupervised learning of topic models to explore the unseen themes with each label, as well as unlabeled themes in the large collection of data. As illustrated in Fig. 3, PLDA assumes that the document's words are drawn from a document-specific mixture of latent topics, where each topic is represented as a distribution over words, and each document can use only those topics that are in a topic class associated with one or more of the document's labels. This approach enables PLDA to detect extensive patterns in language usage correlated with each label.

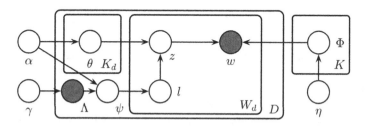

Fig. 3. Probabilistic graphical model for PLDA: each document's word w and label Λ are observed, with the per-doc label distribution ψ, per-doc-label topic distributions θ, and per-topic word distributions Φ hidden variables. Because each document's label-set λ_d is observed, its sparse vector prior γ is unused; included for completeness. η are the parameters of the word prior whereas l is a label and z a topic. For further details please check the work in [21].

4   Experimental Dataset

In this section, we describe the dataset used in our experiments and the applied preprocessing steps. The used dataset consists of a collection of psychotherapeutic transcripts available for research. The transcribed and collected conversations adhere to the guidelines of the American Psychological Association (APA)[4]. An approval to use the collection was granted by an Internal Committee of Biomedical Experiments (ICBE) of Philips after a review of the agreements, the consent procedures, and data handling plan by legal and privacy experts. We remark that meta-data preprocessing has been executed for the two tasks we present in this paper (TDT), whereas text preprocessing has been run on topic detection only.

---

[4] http://www.apa.org.

4.1 Data Description

Counseling and psychotherapy transcripts contained in our data encompass a diverse set of patients, a large-scale array of topics, and different therapeutic strategies. Hundreds practicing counselors worldwide have transcribed and rendered the used conversations according to APA Ethics Guidelines for use and anonymity. The dataset consists of 1729 transcripts of 1:1 conversation with a total of 340,455 talk turns, 75,732 unique terms, and more than 9 million words. Each transcript has on average 200 talk-turns and eight words for talk-turn. They are also extended with meta-data consisting of the corresponding school of psychotherapy, counselors-patients information such as gender, age range, and sexual orientation as well as a table of topics discussed during the therapeutic conversation. Two different kind of information is contained in the table of topics, that is:

- *Subjects*, which are specified hierarchically into three consecutive levels. The top level is the most general subject, whereas the remaing two levels are more precise[5]
- *Symptoms*, which are overall 79 symptoms defined in the DSM-IV[6] manual, like, for example, *Depression, Anger, Fear*. Reason why we have adopted the DSM-IV and not the newer DSM-V is because the dataset we have employed has been structured according to DSM-IV.

4.2 Preprocessing of the Meta-data

Considering the high number of items in the table of topics, similar topics have been merged experimentally by means of the following steps:

1. Eliminate all the subjects and symptoms that occur in less than 3% of the dataset;
2. Group together all the subjects belonging to the same Wikipedia category[7] regardless their position in the given hierarchical structure.
3. Assign a label to the new subject according to the psychology topics table from APA. For example *Parent-child_relationship* and *Family* are mapped to a new subject from APA known as *Parenting*.
4. Reduce the number of symptoms by using the DSM-IV manual with the expert support of a counselor. In particular, we group symptoms with high-level correlation into a representative one. For example, *Sadness* and *Hopelessness* are merged into the symptom: *Depression*.

In this way the final set, illustrated in Fig. 4, has been reduced to 18 subjects and 16 symptoms only.

---

[5] For example, the word *Family* could correspond to a top level topic, while *Family violence* and *Child abuse* would be associated to the second and third levels respectively. Up to 575 subjects have been used in the three levels in total.

[6] https://dsm.psychiatryonline.org.

[7] https://en.wikipedia.org/wiki/Category:Main_topic_classifications.

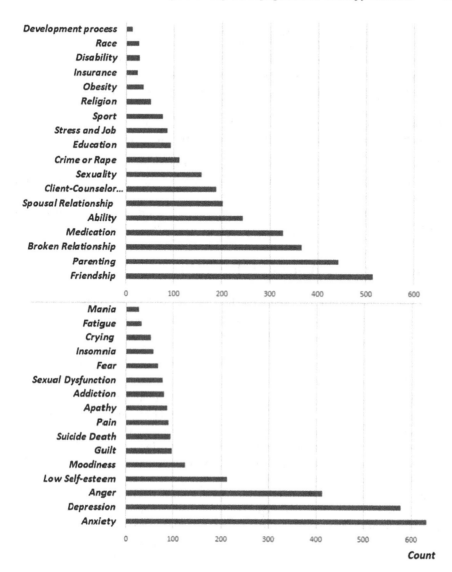

Fig. 4. The resulting 18 subjects (up) and 16 symptoms (bottom)

4.3   Preprocessing of the Conversation Text

A number of classic NLP pre-processing steps [23] have been applied to the dataset by using the NLTK platform[8]. The performed steps include:

1. tokenization, which transforms texts into a sequence of tokens;

---

[8] http://www.nltk.org/.

2. removal of all punctuations, stop words, numbers, words that frequently appeared in the text with minor content information (e.g., "mm-hmm"), and words that occurred in less than five documents whereas keeping nouns, verbs and adjectives only. This step was achieved by using unigram part-of-speech tagging [19], contained in the NLTK framework, in order to identify word types in each talk turn.
3. removal of the most common words (100 overall, in our case), and of the talk turns with one word only or with words shorter than three characters.

Note that the stemming and lemmatization steps have been omitted on purpose to avoid negative impact because the resulting changes may influence the evaluation of the topic model.

After the performed pre-processing steps, the resulting corpus consists of 2,849,457 tokens (14,274 unique ones) and a total of 268,478 talk turns.

5   The Proposed Approach for TDT

As described below, our proposed method consists of three phases; (1) topic modeling, (2) assignment of topic labels to talk turns, and, finally (3) tracking of the propagation of topics over the conversation.

5.1   Topic Modeling

The topic detection was performed using a PLDA implementation based on the Stanford Topic Modeling Toolbox[9](TMT). The model requires a set of parameters including the number of hidden topics to be discovered, the α and η hyperparameters (see Fig. 3), and a training corpus. We define each talk-turn as a document. This results in a total of 268,478 documents after the reprocessing step. Each document is associated with the corresponding topics from the table of topics of the corresponding transcript. Additionally to the 34 predefined topics inferred from the metadata, we experimentally set the number of hidden topics to 20. The hyperparameters α and η are set to 0.01. Based on those parameters, we train our model with 150 epochs using an approximate variant of the collapsed variational Bayes algorithm or the so called VB0 algorithm [4].

5.2   Topic Inference

The learned model delivers a weighted set of words per topic as illustrated in Table 1. The table depicts the top ten terms for each topic. The first column shows an example of a latent (i.e. discovered topic) whereas the second and third columns show two predefined topics together with their related words. As expected, terms associated to a particular topic tend to be semantically related particularly for subjects and symptoms. For example, the topic *Parenting* is defined by terms including family members, such as *mom, mother,*

---

[9] https://nlp.stanford.edu/software/tmt/tmt-0.4/.

dad, etc. whereas the topic *Addiction* includes terms close to alcohol and drugs (*drinking, smoke, etc.*). The same holds for latent topics, where *Topic-5*, for instance, includes similar terms related to *work*. Based on the learned weighted lists of terms, the model infers the most likely topics given a particular talk turn. Figure 5 depicts the *per-document topic distribution* of the five most likely topics in a selected conversation (i.e. *Stress and Job*; *Suicide and Death*; *Sexuality*; *Depression*; *Fear*).

Table 1. An illustrative example of the three kinds of topics and their most likely associated terms. *Topic-5* shows an example of the discovered topic, *Parenting* presents an example of a known subject, and *Addiction* presents an example of a known symptom.

Discovered topic: *Topic-5*		Known subject: *Parenting*		Known symptom: *Addiction*	
Associated words	Weight	Associated words	Weight	Associated words	Weight
Pay	1125386	*Mom*	1354.692	*Drinking*	120.9583
Month	957.7061	*Mother*	1190.696	*Drugs*	78.25677
Working	809.5385	*Dad*	1103.559	*Alcohol*	60.09412
End	799.9208	*Family*	996.8899	*Drug*	59.39509
Help	649.7758	*Brother*	786.8062	*Stoned*	47.30778
Giving	516.6524	*Parents*	745.5274	*Smoking*	44.19726
Months	502.375	*Father*	689.4923	*Marijuana*	41.16274
Year	463.4732	*Sister*	490.4897	*Girlfriend*	37.31469
Paid	421.3631	*Kids*	376.6678	*Smoke*	36.22964
Paying	416.7115	*Children*	313.5798	*Uptight*	35.90694

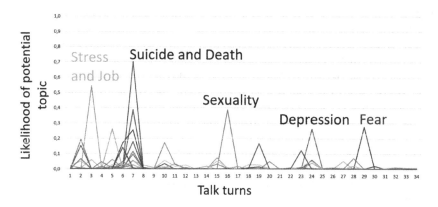

Fig. 5. Example of the *per-document topic distribution* in each talk-turn on a conversation.

Table 2 reports some examples of talk-turns of the patient and their associated representing topics, with the corresponding probabilities, produced by our PLDA-based method.

Table 2. Examples of talk-turns and their associated topics after PLDA within different conversations

Talk-turns	Associated topic
I came into it late, and it was a story about a father and daughter. And it was very much about feelings and...this was a man whose only family was his daughter and...had reappeared in her life and all that. And I remember thinking "Oh I bet Dad's not watching this at all." Or...is not enjoying it because I don't think ever of my family could feel ever be shared. Not with mom and me but even there I mean there were layers of...constraint. Etc.	Parenting, 40%
No. I don't know-maybe I do like it underneath it all. You know it keeps coming back to this climaxing - that I think I would enjoy intercourse if I could climax. And that seems to be you know, know-maybe know keeps coming climaxing think to enjoy intercourse climax seems to know	Sexual dysfunction, 92%
It's a kind of close friendship I guess of being able to just talk to them about anything or to not talk to them about anything. I mean just to sort of be able to be with them and have them understand how your feeling if you happen to be feeling any way at all or do enjoy things with you. Etc.	Friendship, 46%
Right. These feelings. That you know beginning to wonder if you know. You know I'm going to be this unhappy in marriage. To feel this lonely in the marriage. I don't want to be alone you know. In fact, I have all of the responsibility but none of the advantages. I want just to know have some of the advantages of being alone. And it feels pretty screwed up	Spousal relationship, 42%
Like sometimes when I'm thinking about sex or just getting away from everything including the person I'm talking to, and I don't feel like I can say that to a person right to his face	Sexuality, 82%
I never get really happy about anything very rarely, and at the same time, I never get really depressed about anything. I just don't let myself you know. And I was consciously sitting there trying to get - I mean after I started getting depressed I decided to relax and get just as depressed as I could get because Meg says that often helps	Depression, 46%
It's craving the marijuana. It's craving the alcohol. It's craving you know whatever it is	Addiction(s), 99%
All right sure. What effect does the medications we have you on now which is predominantly Lamictal and we have you on some Trazodone at night for sleep and I understand that s a catch 22 type of medication	Medication, 76%

5.3   Tracking Topics

In order to investigate how topics propagate throughout the given therapeutic conversations, we built two topics transition matrices (TTMs). For convenience, we added a new topic annotated as *Meaningless talk* which we associated to talk-turns that provide poor semantics contents or language, or non-verbal communication (e.g. "Yahh!!, Mm-hmm"). In general, the three main types of topics changes are depicted:

1. The counselor keeps talking about the same topic to the patient from the previous talk-turn, and vice versa;
2. The counselor moves to a new topic after the talk-turn of the patient;
3. The patient moves to a new topic after the talk-turn of the counselor.

To illustrate the topics dynamics in more details, we create a patient-to-counselor TTM CP_k that describes all the topic changes within the conversation k. In particular, $CP_k[i, j]$ is the number of times that topic i changes into topic j in a conversation k. We merge the CP_k matrices together by summing them up and obtain our final matrix CP. Similarly, we built a counselor-to-patient matrix PC by using the topics-change defined earlier by switching counselor and patient. The arithmetic difference between the two matrices is illustrated in Fig. 6.

The resulting values are mapped into colors to visually indicate the different levels of engagement between the counselor and the patient: black corresponds to values below -10, grey to values between -10 and $+10$, and greater white for values greater than 10. Thus, the diagonal of the matrix gives an idea about "resistance level" to a particular topic. In general, there is a prevalence of the grey color in the diagonal (17 values) which suggest that the clients are mostly open to continuing the same topic. With twelve black values and six white values, the diagonal also suggests that counselors tend to switch topics twice as the patients do. A possible explanation is that counselors aim at searching for other correlated symptoms or subjects that would lead to a mental disease. The other values of the matrix describe the second and third type of topic changes; the number of white and black values are approximately equal, which means that the conversations, in general, are discussed without perceived tactics. Nevertheless, some rows and columns are mostly either negative or positive (e.g. *Parenting*) which indicate the potential use of some strategies. The counselor often switches the topic if the previous one was *Mania*, *Medication* or *Patient-Counselor Relations*. Instead, he/she frequently starts a new topic if the patient's talk restrains less semantic contents (*Meaningless Talk*). On the other hand, the patient often switches topics if the previous topic was related to *Parenting*, *Friendship*, *Sexual dysfunction*, *Crying*, or *Stress-and-Work*.

6   Evaluation

The evaluation of the performance of a topic model is not an easy task. In most cases, topics need to be manually evaluated by humans, which may express different opinions and annotations. The most common quantitative way to assess

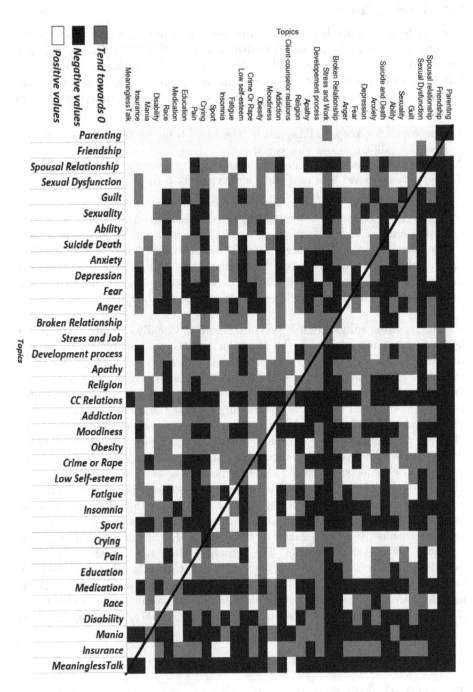

Fig. 6. The difference matrix between CP and PC.

a probabilistic model is to measure the log-likelihood of a held-out test set performing perplexity. However, the authors in [10] have shown that, surprisingly, perplexity and human judgment are often not correlated, and may infer less semantically meaningful topics. A potential solution to this problem is provided by the topic coherences, that is a typical way to assess qualitatively topic models by examining the most likely words in each topic. For such a purpose, we employed Palmetto[10], a tool to compute topic coherence of a given word set with six different methods. The one that we selected for our purposes was the C_V method [22], which uses word co-occurrences from the English Wikipedia, and that has been proven to highly correlate with human ratings. C_V is depended on a one-set segmentation of the top words and a measure that uses normalized pointwise mutual information. The one-set segmentation computes the cosine similarity between each top words vector and the amount of all top words vectors. The coherence value is then the arithmetic average of these similarities and represents an intuitive measure of the goodness of the topics produced by PLDA. In this work, we evaluated our PLDA topic model for topic detection using C_V coherence. In particular, we gave the top five terms (according to the weight of PLDA shown in Table 1) for each of the 34 topics as the input, obtaining as output a satisfactory coherence amongst all the detected topics. Indeed on average a topics coherence value larger than 50% was obtained, which is recognized in the research community already as a well-acceptable coherence score for a TDT model. This further substantiates the validity and potentials of our method.

7   Conclusions

In this paper, we study the topic propagation in a large collection of transcriptions from real psychotherapeutic sessions. For topic modeling, we used Partially Labeled Latent Dirichlet Allocation, PLDA, which makes it possible to track both common topics that are known in advance, and topics encountered in the conversational data. Moreover, we used topic coherence evaluation algorithms to evaluate the consistency of the topic system. Finally, we computed TTM to capture the dynamics of each ongoing topic in the conversations understanding the patterns how the patient and therapist, respectively, maintain and switch topics during the therapy sessions.

Knowing how topics change and propagate over the session can be used by counselors to drive the discussion and to adjust their assessment of the emotional state and barriers of the patient. These aspects of interaction are critical for all mental health specialists as they are related to the health state of the patients. We conclude that PLDA and TTM may be of benefit to the therapeutic conversational speech analysis and other real-life applications of AI to psychotherapy.

---

[10] http://aksw.org/Projects/Palmetto.html.

References

1. Allahyari, M., Kochut, K.: Automatic topic labeling using ontology-based topic models. In: 2015 IEEE 14th International Conference on Machine Learning and Applications (ICMLA), pp. 259–264, December 2015
2. Angus, D., Smith, A.E., Wiles, J.: Human communication as coupled time series: quantifying multi-participant recurrence. IEEE Trans. Audio Speech Lang. Process. **20**(6), 1795–1807 (2012)
3. Angus, D., Watson, B., Smith, A., Gallois, C., Wiles, J.: Visualising conversation structure across time: insights into effective doctor-patient consultations. PloS ONE **7**(6), e38014 (2012)
4. Asuncion, A., Welling, M., Smyth, P., Teh, Y.W.: On smoothing and inference for topic models. In: UAI 2009, pp. 27–34. AUAI Press, Arlington (2009)
5. Atkins, D.C., Steyvers, M., Imel, Z.E., Smyth, P.: Scaling up the evaluation of psychotherapy: evaluating motivational interviewing fidelity via statistical text classification. Implementation Sci.: IS **9**, 49 (2014)
6. Bangalore, S., Di Fabbrizio, G., Stent, A.: Learning the structure of task-driven human-human dialogs. IEEE Trans. Audio Speech Lang. Process. **16**(7), 1249–1259 (2008)
7. Blei, D.M., Lafferty, J.D.: Dynamic topic models. In: ICML 2006, pp. 113–120. ACM, New York (2006)
8. Blei, D.M., Ng, A.Y., Jordan, M.I.: Latent Dirichlet allocation. J. Mach. Learn. Res. **3**, 993–1022 (2003)
9. Breuing, A., Wachsmuth, I.: Talking topically to artificial dialog partners: emulating humanlike topic awareness in a virtual agent. In: Filipe, J., Fred, A. (eds.) ICAART 2012. CCIS, vol. 358, pp. 392–406. Springer, Heidelberg (2013). https://doi.org/10.1007/978-3-642-36907-0_26
10. Chang, J., Boyd-Graber, J., Gerrish, S., Wang, C., Blei, D.M.: Reading tea leaves: how humans interpret topic models. In: NIPS 2009, pp. 288–296. Curran Associates Inc., Red Hook (2009)
11. Chen, W.T., Lin, S.C., Huang, S.L., Chung, Y.S., Chen, K.J.: E-HowNet and automatic construction of a lexical ontology. In: COLING 2010, pp. 45–48. Association for Computational Linguistics, Stroudsburg (2010)
12. Chen, Y., Liu, L.: Development and research of topic detection and tracking. In: 2016 7th IEEE International Conference on Software Engineering and Service Science (ICSESS), pp. 170–173, August 2016
13. Drew, P., Chatwin, J., Collins, S.: Conversation analysis: a method for research into interactions between patients and health-care professionals. Health Expect.: Int. J. Publ. Particip. Health Care Health Policy **4**(1), 58–70 (2001)
14. Gaut, G., Steyvers, M., Imel, Z.E., Atkins, D.C., Smyth, P.: Content coding of psychotherapy transcripts using labeled topic models. IEEE J. Biomed. Health Inform. **21**(2), 476–487 (2017)
15. Gelbukh, A.: Natural language processing. In: Fifth International Conference on Hybrid Intelligent Systems (HIS 2005), pp. 1–6 (2005)
16. Howes, C., Purver, M., McCabe, R.: Investigating topic modelling for therapy dialogue analysis. In: Proceedings of the IWCS 2013 Workshop on Computational Semantics in Clinical Text (CSCT 2013), pp. 7–16. Association for Computational Linguistics (2013)
17. Imel, Z.E., Steyvers, M., Atkins, D.C.: Computational psychotherapy research: scaling up the evaluation of patient-provider interactions. Psychotherapy (Chicago, Ill.) **52**(1), 19–30 (2015)

18. Mohr, J.W., Bogdanov, P.: Introduction-topic models: what they are and why they matter. Poetics **41**(6), 545–569 (2013). Topic Models and the Cultural Sciences

19. Khin, N.P.P., Aung, T.N.: Analyzing tagging accuracy of part-of-speech taggers. In: Zin, T.T., Lin, J.C.-W., Pan, J.-S., Tin, P., Yokota, M. (eds.) GEC 2015. AISC, vol. 388, pp. 347–354. Springer, Cham (2016). https://doi.org/10.1007/978-3-319-23207-2_35

20. Ramage, D., Hall, D., Nallapati, R., Manning, C.D.: Labeled LDA: a supervised topic model for credit attribution in multi-labeled corpora. In: EMNLP 2009, pp. 248–256. Association for Computational Linguistics, Stroudsburg (2009)

21. Ramage, D., Manning, C.D., Dumais, S.: Partially labeled topic models for interpretable text mining. In: KDD 2011, pp. 457–465. ACM, New York (2011)

22. Röder, M., Both, A., Hinneburg, A.: Exploring the space of topic coherence measures. In: WSDM 2015, pp. 399–408. ACM, New York (2015)

23. Uysal, A.K., Gunal, S.: The impact of preprocessing on text classification. Inf. Process. Manag. **50**(1), 104–112 (2014)

24. Yeh, J.F., Tan, Y.S., Lee, C.H.: Topic detection and tracking for conversational content by using conceptual dynamic latent Dirichlet allocation. Neurocomputing **216**(Suppl. C), 310–318 (2016)

Dr. AI, Where Did You Get Your Degree?

Edward Raff[✉], Shannon Lantzy[✉], and Ezekiel J. Maier[✉]

Booz Allen Hamilton, McLean, USA
{raff_edward,lantzy_shannon,maier_ezekiel}@bah.com

Abstract. Federal health agencies are currently developing regulatory strategies for Artificial Intelligence based medical products. Regulatory regimes need to account for the new risks and benefits that come with modern AI, including safety concerns and unique opportunities, like the potential for autonomous learning, that makes AI dramatically different from traditional static medical products. The current default regulatory regime is to treat AI like a medical device (i.e., as opposed to like a drug or a biologic product). As agencies like the U.S. Food and Drug Administration (FDA) develop new regulation to cover the uniqueness of AI, we suggest they consider adopting aspects of regulation traditionally used in the practice of medicine (i.e., doctors). In fact, FDA is currently undergoing a pilot that moves in that direction. We propose that AI regulation in the medical domain can analogously adopt aspects of the models used to regulate *medical providers*. We provide this view point to encourage discussion of how medical AI might be regulated. In doing so, we will also review several issues our framework does not resolve.

Keywords: Regulation · Continuous learning · Clinical applications

1 Introduction

Governmental agencies like the FDA are anticipating a wave of new software products for medical applications, and are currently drafting regulatory guidance in anticipation of this wave. Goals of new regulatory guidance include protecting the public from risk, reducing the time to market for these devices, and fostering an innovative market for the new software. For example, the FDA's Digital Health Program is running an nine-company pilot program[1] to pre-certify organizations developing software as a medical device (SaMD) for streamlined pre-market review [1]. However, FDA's recent draft publication[2] stops short of providing guidance for artificial intelligence as a medical device (AIaMD). In this

---

[1] https://www.fda.gov/medicaldevices/digitalhealth/digitalhealthprecertprogram/default.htm.

[2] https://www.fda.gov/downloads/MedicalDevices/DeviceRegulationandGuidance/GuidanceDocuments/UCM524904.pdf.

The views expressed in this paper come solely from the authors, and do not represent an official or endorsed position by Booz Allen Hamilton.

paper, we extend the SaMD discussion to a regulatory framework for AIaMD. For the purpose of this paper, we define AI-enabled medical device as a software product that actively learns after it is released to the market, and that is intended to inform or make decisions on behalf of a health care provider or patient.

Medical products, such as drugs, biologics, and non-AI devices undergo an evidence-based review of their safety and efficacy, i.e., their benefit-risk profiles. AIaMD will upend this traditional regulatory paradigm because, by definition, the devices can automatically change their own benefit-risk profile without human intervention. For example, an algorithm to detect cancer from MRI imagines and recommend treatment pathway could become more precise and sensitive over time by learning from cases in situ. While we have not yet seen reports of AIaMD in the health care market, it is crucial that governments provide clear guidance on how upcoming AIaMD product submissions will be reviewed and approved. Promising AI-enabled medical products have surfaced, albeit ones that do not continuously learn. For example, in early 2017, Arterys Inc. received FDA 510(k) clearance for its web-based medical imaging analytics software[3]. The lack of AIaMD submissions may be due to lack of sufficient readiness of the technology, but it may also be stymied by the lack of clear regulatory guidance and government approval pathways. The development of clear AIaMD regulation will provide market stability and encourage innovation due to: (1) improved consumer confidence in the safety and efficacy of products; (2) a clear understanding of the requirements for marketing approval, thereby allowing companies to judge risk of their investment going to market, and informing academic and institutional review boards of the requirements surrounding medical studies. As AI researchers, it is critical we have a voice in how this regulation forms to ground expectations and ensure that innovation is not unduly stifled.

We believe there is a risk that harmful regulation could be established (i.e., regulation that does not increase safety and efficacy but prevents or slows innovation) due to fear and the uncertainty around AIaMD. For example, the often "black-box" nature of AI has spurred considerable demand for interpretability and explainability in an AI-based medical device [10]. A "right to explanation" has already been codified in the European Union's laws [7]. Regulatory review of medical products traditionally focuses on evidence of *safety* and *effectiveness* over interpretability or mechanism of action. We contend that *mandating* interpretability is excessively burdensome for AI-enabled devices. This is not to say that interpretability has no value; AI systems that can explain their choices may warrant faster regulator approval. But to focus on interpretability as a necessity for AI would stifle progress.

Rather than focusing regulation on algorithm explainability and self-updating models, we would like to shift focus to outcomes for the patient and to the healthcare market. In this paper we use the paradigm of regulating the practice of medicine as a framework for thinking differently. We propose elements of a framework analogous to the standards used to license medical providers.

---

[3] https://www.forbes.com/sites/bernardmarr/2017/01/20/first-fda-approval-for-clinical-cloud-based-deep-learning-in-healthcare.

Similar to accredited *medical schools* which train medical doctors, we contend that AI-enabled devices should be trained utilizing accredited *data collection and validation methods*. AIaMD trained using these accredited data collection and validation methods should then be evaluated based on measured outcomes to individual patients. Similar to state medical boards which remove harmful doctors from practice, we contend that we need an AI regulator to surveil and remove AIaMD if they become harmful.

2   Regulatory Design for AIaMD

To ensure that the immense potential of AI is not hampered, stakeholders must actively engage in the development of the regulatory framework. Researchers, software product developers, patient advocates, medical providers, and payers' participation in this discussion will help to avoid the hype and fear that has led to previous AI winters. We argue that the methodological accreditation and outcomes-focus framework outlined below, will enable regulatory agencies to accomplish their mandate of protecting public health while allowing for innovation by AI researchers. However, discussion, dialogue, and iteration is needed. The FDA has invited public feedback and participation in the conversation.[4]

2.1   Accrediting Our Data Sources and Methods

Doctors are educated by accredited universities. AIaMD should be trained with accredited data and methods. While much of the discussion around AI focuses on the algorithms used, data collection and the training methods are extremely important to the success of any model. AI is not immune to the "garbage-in garbage-out" problem, and so ensuring that high-quality algorithms are developed means we must ensure data is of an equally high quality. Accrediting the process by which data is acquired and prepared provides the foundation needed for any level of trust in the results. Accreditation of a dataset's labeling and creation process should mirror the acceptance criteria of sufficient evidence for new clinical guidance in medical practice. For example, the dataset accreditation scheme should consider: an appropriate diversity of patient backgrounds (e.g., age, BMI, etc); a diversity of feature sources (e.g., MRI images used for training must come from multiple MRI machines of differing versions and differing vendors); the consistency of feature sources between the training and clinical contexts; the completeness of data meta-information; defined measurable and clinically-relevant outcomes (e.g., real-time insulin levels), rather than measures that may be available (e.g., unqualified claims records). Fully satisfying all of these goals may not be possible in each case, but should always be considered and addressed. Significant failures in any of these sub-components can prevent development of actionable and effective AI solutions. For example, [12] found

---

[4] https://www.fda.gov/MedicalDevices/DigitalHealth/DigitalHealthPreCertPro gram/default.htm\#getinvolved.

that out of 2,511 recent genome-wide association studies, 81% of all participants were of European ancestry. This poses a risk that developed solutions and results will be ineffective for the majority of the world's population.

As part of Booz Allen Hamilton's organization of the 2016 and 2017 Data Science Bowl competitions [2], which focused on detecting heart function and lung cancer respectively, organizers examined each of these aspects of the competition data to ensure that it was high-quality and enabled the development of useful algorithms. We found unexpected metadata which artificially boosted the algorithm's appearance of clinical performance (i.e., leakage). Specifically, meta-information describing the hospital that labeled the cardiac MRI images proved to be strongly predictive of a specific heart measurement, despite having no clinical diagnostic power. If this meta-information was not recorded, organizers would not have discovered the correlated, but not actionable feature, and could have led to model overfitting to the training data. This exemplifies why data should be acquired from a diversity of locations, and why trained medical providers must be part of the data preparation process. As one step toward ensuring the safety, AI-enabled devices must be robust to a diversity of input sources. The best way to achieve this robustness is to utilize a diverse high-quality data set for training.

It is possible for regulators to take a proactive approach by creating gold-standard data sets for important and prevalent conditions. Such data could be used in multiple ways to both improve the efficiency of regulation and the speed at which products are developed. These could be kept as secret evaluation sets to confirm reported performance, an independent training set to independently test system generalizability, or even provided to product developers to reduce data acquisition costs and promote marketplace competition. The FDA is already exploring the development and curation of a standard dataset for radiogenomics [8]. This could also allow the FDA to preemptively remove barriers that slowed the adoption of Electronic Medical Records in the United States relative to other nations, such as lack of capital and standardized data exchange formats [3].

2.2 Focus on the Outcomes

Doctors' outcomes are monitored by their medical boards, colleagues, and patients; AIaMD postmarket surveillance should include a diversity of feedback sources. By definition, AIaMD learn from well-defined outcomes which are measured while in use. Therefore, post-market surveillance (i.e., monitoring the benefit-risk profile of a medical product after it has been released on the market) can be built directly into an AI product. AIaMD developers should focus on building a system capable of collecting the right outcomes. Regulators should focus on the process by which an AI device developer defines, collects, and uses post-market outcomes to refine and improve the model. Next, similar to a doctor who is subject to review and possible sanctions by their state medical board (i.e., probation periods with added surveillance, or suspension from medical practice), regulators should sanction and/or withdraw an AIaMD from the market for egregious errors.

We propose that, like medical review boards for medical providers, regulators should institute AI review boards consisting of a multidisciplinary group of experts from within and outside the regulatory agency. The AI boards would include continuing education-like requirements to update AI models using new standards and ground truth data, sanctioning AI producers for errors or AI misconduct or bias, and removal of an AI product when it does harm. Trials and studies will remain necessary to ensure that the device is both safe (does no harm), and effective (provides meaningful and quantifiable improvement in outcomes).

3   So Can We Treat AI Like a Doctor?

Framing the regulation of AI in the same manner as medical doctors provides a basis for constructing regulation for non-static products. This approach allows regulators, the AI community, and the general public to debate the opportunities and obstacles of AI-enabled medical devices.

A primary psychological benefit of this approach is to avoid the problem of moving goal posts or an AI double standard. The public is often unwilling to trust a machine to perform a task unless the outcome is *far better* than what a human can produce.[5] This thought process ignores the intrinsic benefits of availability and faster decision making. For example, AI-enabled medical devices can provide both routine care in rural and poor communities that would have no access otherwise, and faster diagnosis, leading to improved patient outcomes. With regulation focused on data accreditation and clinical outcomes, regulators avoid unnecessarily delaying adoption of AI technology for medicine.

This regulatory framework also provides guidance on ensuring AI devices remain safe over time. Physicians are not simply told to do no harm. Rather, physicians progress from interns to specialist over their careers, and as they progress their responsibilities and autonomy increases. AI devices could follow a similar (task-dependent) progression. This lends to a natural encouragement for AI products to be developed in an incremental approach. However, AI devices need not progress completely to autonomous continually learning agents (i.e., a specialist). Instead AI devices can ultimately be tools, which have utility to physicians irrespective of their autonomous continually learning capability.

With this regulatory approach we must collectively recognize that errors and mistakes will be made. Just as doctors, drugs, and devices sometimes unintentionally harm patient, AIaMD will as well. Just as deaths due to medical errors occur, so do deaths caused by software bugs [9]. Every death is tragic; yet the question of safety is not whether a doctor or an AIaMD prevents all harm, but rather he/she/it reduces the rate of harm from the current standard of care. SaMD deaths in Leveson [9] were incidents that the FDA studied in order to remediate and prevent future incidents. While the hope for AI devices is to reduce the frequency of such unfortunate incidents, the same lessons will apply to the AI space. Researchers who acquire and prepare the data, and develop

---

[5] https://phys.org/news/2016-05-humans-automated-advisor-bad-advice.html#jCp.

models to analyze take action must understand this risk. Given the potential greater autonomy of AIaMD, AI developers may require a form of "malpractice" insurance. This insurance would provide fiscal and regulatory incentives to encourage safety and provide financial recompense when incidents occur.

4   Failure Points

We believe lifting and adapting from the regulatory framework for medical providers is useful to frame our discussion around regulating AI. However, that framework is not perfect as it exists today, and we see no reason to expect it will be perfect for AI either. It is important to also discuss points where the regulatory schema for the practice of medicine will not work for AIaMD. In the sections below we discuss these failure points and offer prompts to develop thought and discussion from the community. Below we will discuss three issues, which we feel are important toward developing complete regulation.

4.1   Recalling AI

The reach of bad AIaMD will be broader than the reach of a bad doctor. Every year thousands of doctors are sanctioned by their state medical boards. Morrison and Wickersham [11] found that 79% of California's disciplinary cases resulted in some form of license suspension or revocation. This is an important issue and part of the reason physicians are licensed, but it is also reactive—action does not occur until something goes wrong. During the time between misconduct and revocation, these doctors are unfortunately putting their patients at risk. Similarly, some AIaMD products will need to be recalled (i.e., have their "license" suspended) in the same reactive manner. We will again have an issue with the time between product failure ("misconduct") and a successful removal from the market. But in this case, an AI product could have potentially been deployed nation wide or even globally, where a single doctor's misconduct is intrinsically limited to a smaller pool of people. This increases the potential cost (e.g., of patient well being, potential monetary damages) of an AI failure case.

AIaMD may be less fungible than individual doctors, making removal more disruptive. Removing AIaMD may also be more locally disruptive than removing a bad doctor because it may be too unique. If one doctor is removed from medical practice, there are other doctors who can step in to perform the functions. However, if AIaMD performs a unique function that becomes an essential part of a clinical workflow, it may be more difficult to replace the function. For example, if radiologists begin to rely heavily on computer analysis of tumor images, removing that AIaMD may cause a temporary lapse in care for tumor analysis.

4.2   Adversarial AI and Security

While the medical industry has long had to handle sensitive personally identifiable and protected health information, security of this information has not historically been reviewed by regulatory agencies. The Health Insurance Portability and Accountability Act (HIPAA) laws in the U.S. provide some regulation regarding security issues, and will require updates as AIaMDs enters the market. With the advent of adversarial machine learning, a new kind of security issue must also be considered.

In adversarial machine learning, a hypothetical adversary attempts to trick a classifier into making specific incorrect decisions [5]. This research field, which is at the intersection of machine learning and computer security, is most prevalent in the fields of spam filtering, malware detection, and computer vision community, including for self-driving cars. Due to the potential to interact with adversaries, AI-enabled medical device developers must also consider this form of attack. Notably, AI-enabled medical device adversarial interactions may be with individuals engaging in drug-seeking behaviour, as well as sophisticated malicious groups. Fraud is already an enormous issue in the medical field, with hundreds of billions of dollars lost, and there is fear that this problem will only worsen with the adoption of machine learning systems [6].

Ultimately, it is not yet known to what degree adversarial attacks will affect SaMD and AIaMD. AIaMD developers can follow current practices of defining a threat-model by which adversaries can act to evaluate the risk to their systems [4]. However, it has so far been found that such attacks are easy to create and apply, even with threat-models that are highly restrictive to the adversary's actions and knowledge [5]. Regulators must eventually decide how far AIaMD developers must go to protect systems from attacks, and determine in advance domains where their product should not be applied due to risk of attack. This is an issue that will require careful consideration, and by its very nature, not one that we can rely on current systems to handle.

5   Conclusion

Fundamentally, medical regulation exists precisely because without it consumers cannot reasonably assess the quality of all possible medical diagnoses and the benefits and risks of recommended treatments. Regulatory agencies are developing new policy and guidance for static SaMD, and will soon codify rules to govern dynamic AIaMD. Rather than developing new regulations based on our existing rules for static medical products, we proposed using the analogy of medical practice regulation as a foundation to develop a novel regulatory framework for AI-enabled devices. We argue that the regulatory framework for medical practice provides a natural paradigm to address the public's concerns about the use of AI in healthcare, and we have used it to illustrate points of consideration for new regulation. Though the accreditation process for medical doctors is not perfect, the approach has served society for decades and can serve as the foundation for regulating AI-enabled medical devices.

Acknowledgments. We thank anonymous reviewers as well as participants in AIH 2018 workshop at FAIM for their generous feedback and discussions, which have improved this paper.

References

1. Digital health innovation action plan introduction. Technical report, Food & Drug Administration (2017). https://www.fda.gov/downloads/MedicalDevices/DigitalHealth/UCM568735.pdf
2. Data Science Bowl (2018). https://datasciencebowl.com/
3. Anderson, J.G.: Social, ethical and legal barriers to e-health. Int. J. Med. Inform. **76**(5), 480–483 (2007). https://doi.org/10.1016/j.ijmedinf.2006.09.016. http://www.sciencedirect.com/science/article/pii/S1386505606002218. ISSN 1386-5056
4. Biggio, B., Fumera, G., Roli, F.: Security evaluation of pattern classifiers under attack. IEEE Trans. Knowl. Data Eng. **26**(4), 984–996 (2014). https://doi.org/10.1109/TKDE.2013.57. ISSN 10414347
5. Biggio, B., Roli, F.: Wild Patterns: Ten Years After the Rise of Adversarial Machine Learning, pp. 32–37 (2017). http://arxiv.org/abs/1712.03141
6. Finlayson, S.G., Kohane, I.S., Beam, A.L.: Adversarial Attacks Against Medical Deep Learning Systems (2018). https://arxiv.org/abs/1804.05296
7. Goodman, B., Flaxman, S.: European union regulations on algorithmic decision-making and a "right to explanation". AI Mag. **38**(3), 50 (2017). https://doi.org/10.1609/aimag.v38i3.2741. http://www.aaai.org/ojs/index.php/aimagazine/article/view/2741. ISSN 0738-4602
8. Gottlieb, S.: FDA's comprehensive effort to advance new innovations: initiatives to modernize for innovation. Technical report, Food and Drug Administration (2018). https://blogs.fda.gov/fdavoice/index.php/2018/08/fdas-comprehensive-effort-to-advance-new-innovations-initiatives-to-modernize-for-innovation/
9. Leveson, N., Turner, C.: An investigation of the Therac-25 accidents. Computer **26**(7), 18–41 (1993). https://doi.org/10.1109/MC.1993.274940. http://ieeexplore.ieee.org/document/274940/. ISSN 0018-9162
10. Lipton, Z.C.: The doctor just won't accept that! In: Interpretable ML Symposium at NIPS (2017). http://arxiv.org/abs/1711.08037
11. Morrison, J., Wickersham, P.: Physicians disciplined by a state medical board. JAMA **279**(23), 1889 (1998). https://doi.org/10.1001/jama.279.23.1889. http://jama.jamanetwork.com/article.aspx?doi=10.1001/jama.279.23.1889. ISSN 0098-7484
12. Popejoy, A.B., Fullerton, S.M.: Genomics is failing on diversity. Nature **538**(7624), 161–164 (2016). https://doi.org/10.1038/538161a. http://www.nature.com/doifinder/10.1038/538161a. ISSN 0028-0836

Design Principles and Action Reflection for Agent-Based Assistive Technology

Esteban Guerrero[1]([✉])[iD], Ming-Hsin Lu[2][iD], Hsiu-Ping Yueh[2,3][iD],
and Helena Lindgren[1][iD]

[1] Computing Science Department, Umeå University, Umeå, Sweden
{esteban,helena}@cs.umu.se
[2] Department of Bio-Industry Communication and Development,
National Taiwan University, Taipei, Taiwan
{f01630002,yueh}@ntu.edu.tw
[3] Department of Psychology, National Taiwan University,
Taipei, Taiwan

Abstract. This paper is aimed at formalizing the interplay among a person to be assisted, an assistive agent-based software, and a caregiver. We propose general principles for designing the interplay between a person to be assisted and an agent based on formal argumentation theory to characterize the agent's reasoning processes. These principles emerge from a novel perspective to understand assistive technology using the concept of zone of proximal development (ZPD) from social sciences. ZPD can be understood as a measurement of activity development, comparing what a person can perform with or without external help. We characterize a rational agent in four ZPD zones: (I) independent activity execution, agent takes no action; (II) ZPD_H: a person supported by another person, agent takes no action; (III) ZPD_S: a person is supported by an agent; and (IV) ZPD_{H+S}: a person is supported by a caregiver and a software agent at the same time. An algorithm was developed for the agent to reason about the actions to be selected in different situations, based on formal argumentation theory for allowing non-monotonic reasoning. The formal models and algorithm were implemented in a prototype system using augmented reality as interface. Future work includes evaluating the principles and algorithm in actual use situations.

Keywords: Argumentation theory · Rational agents ·
Assistive technology · Human activity · Activity theory

1 Introduction

Intelligent *assistive technology* (AT) is an umbrella of artificial intelligence-based machinery, that in general, is able to observe and reason about appropriate and tailored support to individuals [24]. An AT may have different aims, as an *assurance system*, *compensation system* or as an *assessment system* [31]. No matter what the *assistive goal* is, the internal machinery of an intelligent AT should

© Springer Nature Switzerland AG 2019
F. Koch et al. (Eds.): AIH 2018, LNAI 11326, pp. 84–98, 2019.
https://doi.org/10.1007/978-3-030-12738-1_7

reason about: (1) the client-caregiver interaction[1]; and (2) the context where the AT service is provided. Moreover, the AT system should generate consistent services, as *outputs* of the intelligent system. In contrast to AT provision based on artificial intelligence, public AT service provision is regulated by policies, procedures and approaches, being part of different national or regional health care and welfare systems [39].

In the *deductive systems*[26] literature as part of AI, different effort have been developed to provide formal principles of a deductive system (see [2,7,12, 16]). In this setting, and inspired by public AT efforts to provide high quality services, we propose in this paper a novel formal set of principles that AT based on AI should follow to warranty consistent AT services. We hypothesize that the internal reasoning process of an AT needs to fulfill general principles of consistency and soundness, aiming at not interfere, contradict or disregard client and/or caregiver actions.

To this end, this paper has a two-fold goals: (1) propose a general decision-making mechanism (algorithm) considering information of a client and a caregiver who supports during activity execution; and (2) introduce general principles of no-contradiction to which any AT reasoning about observations, goals and actions of individuals must comply. We framed those principles in a client-caregiver-agent[2] interaction in four common AT scenarios, as follows:

S1. Independent activity execution: a client does not need to be assisted during the execution of an activity, an AT is present but it takes no action.
S2. Human support: a caregiver assists a client that needs support during an activity execution. An AT is present but it takes no action.
S3. Agent supporting: a client is supported by an AT. A caregiver is not present during such interaction.
S4. Joint assistance: a client is supported by a caregiver and an AT.

Formal argumentation theory [3] is used to embed non-monotonic reasoning in an agent, *i.e.*, resembling the kind of assessment reasoning performed by clinicians: (1) gathering data through observations; (2) handling ambiguous and uncertain observation information; (3) generating current function status hypothesis; (4) deduce an explanatory outcome of explanation; and (5) retracting the explanation under new evidence [18]. In this sense, the proposed argumentation-based algorithm (Algorithm 1) takes different decisions depending on the observations, goals and actions in particular scenarios (S1–4).

Scenarios S1–4 are analyzed from an *activity theory* [13,21] perspective, which investigates these AT contexts as a continuum of support adaptation. We analyze S1–4 as "distances" from what a client can do independently, to the activity potential of that client supported by a caregiver or an AT system. We explore

---

[1] In this context, *client* is an individual that receives support from a caregiver and/or an intelligent AT system.

[2] In this paper, an *agent* is an AT machinery based on the concept of software agents that takes decisions about how to support an individual during activity execution see [14].

computational versions of the so-called *zone of proximal development* (ZPD) [38] for each scenario.

We present a basic architecture for an AT system able to identify these assistive scenarios. We implement a prototype of such architecture using a projected *augmented reality* the AT system supports a client displaying personalized information. Our AT system captures information from the client and caregiver using 3D cameras, goals and hypothetical actions are embedded in a program using a multi-agent system platform.

This paper is organized as follows: in Sect. 2 methods and theories utilized as foundation of our proposal are presented. Section 3 introduces an algorithm goal-based reflection as an internal mechanism of an agent. Section 4 introduces a set of general principles that an AT system should fulfill. In Sect. 5, the architecture of an AT system that we developed using projected augmented reality is presented. A discussion with our future paths of this investigation are presented in Sect. 6.

2   Theoretical Background

In this section, some concepts of activity theory [22] and formal argumentation theory [3] are introduced. The former, is used in this paper as a framework to represent knowledge about an activity; the later is used to characterize the internal decision-making process of the intelligent assistive system.

2.1   Activity Theory

In this paper, activity theory is used for two purposes: (1) for knowledge representation, structuring information of clients and caregivers following a hierarchical model; and (2) to understand the potential level of activity achievement of a person.

Activity theory describes an *activity* as a hierarchical structure composed of *actions*, which are composed of *operations* as is represented in Fig. 1. Actions are directed to goals; goals are conscious, *i.e.,* a human agent is aware of goals to attain. Actions, in their turn, can also be decomposed into lower-level units of activity called *operations*. Operations are routine processes providing an adjustment of an action to the ongoing situation, they are oriented toward the *conditions* under which the agent is trying to attain a goal.

In this paper we use *logic programs* to capture information about an activity, we denote P as a program and $\mathcal{L}_P$ the set of *atoms* which appear in such program. In this regard, an *activity model* ($\mathcal{A}$) corresponds to information characterizing mental states of an agent framed on a particular activity. $\mathcal{A}$ can be expressed using propositional logic as a syntax language.

Definition 1 (Activity model). *Let P be a logic program capturing the behavior rules of an activity. An activity model $\mathcal{A}$ is a tuple of the form $\langle Ax, Go, Op \rangle$ in which:*

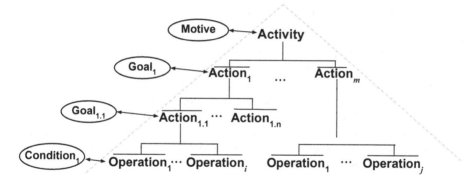

Fig. 1. Hierarchical structure of activity. Activities are composed of actions, which are, in turn, composed of operations. These three levels correspond, respectively, to the motive, goals, and conditions, as indicated by bidirectional arrows. Adapted from [21]

- $Ax = \{ax_1, \ldots, ax_j\}(j > 0)$ *is a set of atoms such that* $Ax \subseteq \mathcal{L}_P$. Ax *denotes the set of actions in* $\mathcal{A}$.
- $Go = \{g_1, \ldots, g_k\}(k > 0)$ *is a set of atoms such that* $Go \subseteq \mathcal{L}_P$. Go *denotes the set of goals in* $\mathcal{A}$.
- $Op = \{o_1, \ldots, o_l\}(l > 0)$ *is a set of atoms such that* $Op \subseteq \mathcal{L}_P$. Op *denotes the set of operations in* $\mathcal{A}$.

In our approach, an activity model $\mathcal{A}$ (Definition 1) may capture information from a client or a caregiver (as in [18]) or/and a software agent-based system (as in [14]). In this paper, we denote $\mathcal{A}_c$, $\mathcal{A}_g$ and $\mathcal{A}_a$ to represent the activity models of a client a caregiver and an agent respectively. In terms of activity theory, $\mathcal{A} = \langle Ax, Go, Op \rangle$ can be seen as a partial description of a complex activity.

In this paper, activity theory is also used to quantify the potential level of activity achievement aiming to frame the decision-making process of the intelligent assistive system. Vygotsky [38] proposed to measure the level of development not through the level of current performance, but through the difference ("the distance") between two performance indicators: (1) an indicator of independent problem solving, and (2) an indicator of problem solving in a situation in which the individual is provided with support from other people [21]. This indicator was coined as a *zone of proximal development* (ZPD) and it has been used extensively in social sciences (see [1,9,19,34]) to understand changes of individuals during assisted learning processes.

In order to create a computable version of the concept of zone of proximal development, we use a function dist that compares two variables (*e.g.* observations of an activity) and returns a numerical value $\alpha \in \mathbb{R}$ representing in this case, a ZPD difference. For convenience, we rename scenarios described in Sect. 1 S1–4 as ZPD_i, ZPD_h, ZPD_s and ZPD_{h+s} respectively.

2.2   Formal Argumentation Theory

Generally speaking, a formal argumentation process can be seen as a mechanism consisting of the following steps (see Fig. 2): (1) Constructing *arguments* (in favor/against a "statement") from a knowledge base; (2) Determining the different *conflicts* among the arguments; (3) Evaluating the *acceptability* of the different arguments; and (4) Concluding, or defining the *justified conclusions*. From artificial intelligence perspective, the important and distinctive characteristics of this process are: (1) their *non-monotonic* behavior, *i.e.*, changing the conclusion when more knowledge is added, and (2) their *traceability*, providing explanations in every step of the reasoning process.

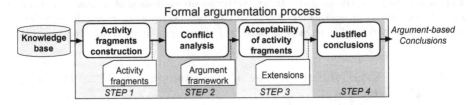

Fig. 2. Inference of an argument-based conclusion using a formal argumentation process

We define the concept of an *activity framework* which frames the necessary knowledge that an agent needs to take a decision.

Definition 2 (Activity framework). *An activity framework ActF is a tuple of the form* $\langle P, \mathcal{H}_A, \mathcal{G}, \mathcal{O}, \mathcal{A} \rangle$ *in which:*

- *P is a logic program.* $\mathcal{L}_P$ *denotes the set of atoms which appear in P.*
- $\mathcal{H}_A = \{h_1, \ldots, h_i\}$ *is a set of atoms such that* $\mathcal{H}_A \subseteq \mathcal{L}_P$. $\mathcal{H}_A$ *denotes the set of hypothetical actions which an agent can perform in a world.*
- $\mathcal{G} = \{g_1, \ldots, g_j\}$ *is a set of atoms such that* $\mathcal{G} \subseteq \mathcal{L}_P$. $\mathcal{G}$ *denotes a set of goals of an agent.*
- $\mathcal{O} = \{o_1, \ldots, o_k\}$ *is a set of atoms such that* $\mathcal{O} \subseteq \mathcal{L}_P$. $\mathcal{O}$ *denotes a set of world observations of an agent.*
- $\mathcal{A}$ *is an activity model of the form:* $\langle Ax, Go, Op \rangle$, *following Definition 1.*

ActF according to Definition 2 defines the space of knowledge of an assistive agent. In this space, an argument-based process (see Fig. 2) can be performed to obtain sets of explainable structures *support-conclusion* for what is the best assistive action to take. These structures can be seen as *fragments* of an activity [18] (see Fig. 3) and can be generated as follows:

Definition 3 (Hypothetical fragments). *Let* $ActF = \langle P, \mathcal{H}_A, \mathcal{G}, \mathcal{O}, \mathcal{A} \rangle$ *be an activity framework. A hypothetical fragment of an activity is of the form* $HF = \langle S, O', h, g \rangle$ *such that:*

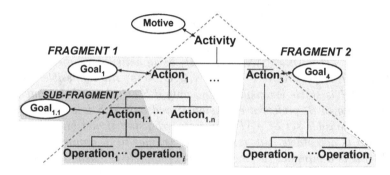

Fig. 3. Fragments and sub-fragments of an hierarchical activity

- $S \subseteq P$, $O' \subseteq O$, $h \in \mathcal{H}_A$, $g \in \mathcal{G}$;
- $S \cup O' \cup \{h\}$ is consistent;
- $g \neq \bot$; and
- S and O' are minimal w.r.t. set inclusion.

Let us introduce a function $\mathsf{Supp}(HF)$ which retrieves the set $\{S, O', h\}$ of a given fragment, which can be seen as the *support* for concluding a goal g. Next step in the argumentation-based process is find different types of contradictions among such fragments (Definition 3): (1) when two fragments have conclusive evidence about opposed achievement of goals; and (2) when a fragment contradicts the support evidence of another. These two types of relationships among fragments resembles the well-known notions of *undercut* and *rebut* in argumentation theory [4, 32].

Definition 4 (Contradictory relationships among fragments). *Let* $ActF = \langle P, \mathcal{H}_A, \mathcal{G}, \mathcal{O}, Acts \rangle$ *be an activity framework. Let* $HF_1 = \langle S_1, O'_1, a_1, g_1 \rangle$, $HF_2 = \langle S_2, O'_2, a_2, g_2 \rangle$ *be two fragments such that* $HF_1, HF_2 \in \mathcal{HF}$. HF_1 *attacks* HF_2 *if one of the following conditions hold: (1)* $g_2 = \neg g_1$; *and (2)* $g_2 \subseteq \mathsf{Supp}(HF_1) = \bot$ *or* $g_1 \subseteq \mathsf{Supp}(HF_2) = \bot$.

An *argumentation framework* is a pair $\langle Args, att \rangle$ in which $Args$ is a finite set of arguments and $att \subseteq Args \times Args$. In [17] an argumentation-based activity framework for reasoning about activities was proposed. We reuse this concept for in our paper, as follows:

Definition 5 (Activity argumentation framework). *Let* $ActF$ *be an activity framework of the form* $\langle P, \mathcal{H}_A, \mathcal{G}, \mathcal{O}, Acts \rangle$; *let* $\mathcal{HF}$ *be the set of fragments w.r.t.* $ActF$ *and* $Att_{\mathcal{HF}}$ *or simply* Att *the set of all the attacks among* $\mathcal{HF}$. *An activity argumentation framework* AAF *with respect to* $ActF$ *is of the form:* $AAF = \langle ActF, \mathcal{HF}, Att \rangle$.

Dung [11], introduced a set of *patterns of selection* of arguments called *argumentation semantics* (SEM)[3]. *SEM* is a formal method to identify conflict

---

[3] Let $SEM()$ be a function returning a set of extensions, given an argumentation framework such as an AAF.

outcomes from argumentation frameworks, such as an activity argumentation framework.

Definition 6. *Let $AAF = \langle ActF, \mathcal{HF}, Att \rangle$ be an activity argumentation framework AAF with respect to $ActF = \langle P, \mathcal{H}_A, \mathcal{G}, \mathcal{O}, Acts \rangle$ An admissible set of fragments $S \subseteq \mathcal{HF}$ is stable extension if and only if S attacks each argument which does not belong to S. preferred extension if and only if S is a maximal (w.r.t. inclusion) admissible set of AAF. complete extension if and only if each argument, which is acceptable with respect to S, belongs to S. grounded extension if and only if it is a minimal (w.r.t. inclusion) complete extension. ideal extension if and only if it is contained in every preferred set of AAF.*

The sets of arguments suggested by SEM are called *extensions*. We can denote $SEM(AAF) = \{Ext_1, \ldots, Ext_k\}$ as the set of k extensions generated by SEM w.r.t. an activity argumentation framework AAF. In this setting, from the perspective of an intelligent agent what it is expected to have is: (1) no contradictory or conflicting sets of fragments sets explaining what is happening in the ongoing activity, and (2) fragments sets defending/supporting a hypothesis about the activity from other fragments. These two notions defines two main concepts in Dung's argumentation semantics: acceptable and admissible arguments.

Definition 7. *(1) An fragment $HF_A \in \mathcal{HF}$ is acceptable w.r.t. a set S of fragments iff for each fragment $HF_B \in \mathcal{HF}$: if HF_B attacks HF_A, then HF_B is attacked by S. (2) conflict-free set of fragments S in an activity is admissible iff each fragment in S is acceptable w.r.t. S.*

Using these notions of fragment admissibility, different argumentation semantics can draw given an activity argumentation framework:

Definition 8. *Let $AAF = \langle ActF, \mathcal{HF}, Att \rangle$ be an activity argumentation framework following Definition 5. An admissible set of fragments $S \subseteq \mathcal{HF}$ is: (1) stable if and only if S attacks each fragment which does not belong to S; (2) preferred if and only if S is a maximal (w.r.t. inclusion) admissible set of AAF; (3) complete if and only if each fragment, which is acceptable with respect to S, belongs to S; and (4) the grounded extension of AAF if and only if S is the minimal (w.r.t. inclusion) complete extension of AAF.*

Conclusions of an argument-based reasoning about an activity may be obtained using a *skeptical* perspective, *i.e.*, accepting only irrefutable conclusions as follows:

Definition 9 (Justified conclusions). *Let P be an extended logic program, $AF_P = \langle Arg_P, At(Arg_P) \rangle$ be the resulting argumentation framework from P and SEM_{Arg} be an argumentation semantics. If $SEM_{Arg}(AF_P) = \{E_1, \ldots, E_n\}(n \geq 1)$, then $\mathsf{Concs}(E_i) = \{\mathsf{Conc}(A) \mid A \in E_i\}(1 \leq i \leq n)$. $\mathsf{Output} = \bigcap_{i=1 \ldots n} \mathsf{Concs}(E_i)$.*
Where E_i are sets of fragments called extensions. The set of all the extensions generated by $SEM_{Arg}(AF_P)$ are denoted as $\mathcal{E}$.

3   Reflection on Decisions About Human Activity

Reflection, as an internal mechanism of a rational agent to (re)consider the best decision alternative (inferring strategies), has been an important line of research in AI particularly in practical reasoning (see [20, 37]). In this paper, we do not consider an agent with *pro-attitudes* as in Bratman model [6], we propose a *control loop* algorithm (as in [33, 36]) to design the action selection and its reflection based on an activity model.

Algorithm 1. Goal-based action reflection

---

input : $\mathcal{E}$ sets of extensions
output: $h \in \mathcal{H}_A$

1 $H \longleftarrow \emptyset$ // list of agent's decisions
2 $\mathsf{Go} \longleftarrow \emptyset$ // list of human's goals
3 $\mathsf{Ref} \longleftarrow \emptyset$ // list of human's reference goals
4 $numExt = |\mathcal{E}|$ // number of extensions
5 $numArg = |E_i|, \ E_i \in \mathcal{E}$ // number of hypoth. decisions per extension
6 $\alpha \longleftarrow 0$ // numeric value of a distance $(0 \leq \alpha \leq 4)$
7 $decisionLat < \alpha, h >=$ // lattice of decisions

8 **for** $i \leftarrow 0$ **to** $numExt$ **do**
9   **for** $j \longleftarrow 0$ **to** $numArg$ **do**
10        $h \longleftarrow \mathsf{Act}\ (hf_j)$
11        $\mathcal{O} \longleftarrow \mathsf{Obs}\ (hf_j)$
12        $\alpha \longleftarrow \mathsf{dist}(\mathcal{O}_{\mathsf{Go}}, \mathsf{Ref}_{\mathsf{Go}})$ // distance function considering
         observations and a reference value *w.r.t.* person goals Go
13        $decisionLat \longleftarrow (\alpha, h)_{hf_j}$ // decision tuple is a ZPD metric for
         the current activity fragment
14   **end**
15 **end**
16 **return** $max(\alpha, h)$

Given a set of hypothetical fragments suggested by an argumentation process, our algorithm selects an agent's action that maximize humans' goals. This mechanism is summarized in Algorithm 1.

Algorithm 1 prioritizes the activity model of a client over an agent and, at the same time, it computes a distance between activity variables. In lines 8–15 of Algorithm 1, such distance is calculated (line 12) over sets of hypothetical fragments. This distance calculation is based on computing a similarity function between the current achievement of human goals in the activity model *w.r.t.* a set of goal reference ($\mathsf{Ref}_{\mathsf{Go}}$ line 12). The dist function in line 12 follows the notion of ZPD, by measuring in every computation the distance between the current development of a person and a reference, which can be given by a caregiver. This approach for comparing current activity execution with a reference has been used in previous approaches [17, 18].

The importance of Algorithm 1 lies on the mechanism for associating a human activity quantification with the internal action decision of an agent. The Algorithm output depends entirely on previous extensions computation. Propositions 1 and 2 present two special cases of agent's behavior when Algorithm 1 is used[4]. One is the possibility to have a conclusion with no action, and the second expresses an inconclusive behavior given that stable semantics may return $\emptyset$ as output.

Proposition 1. *An agent calculating the goal-based action reflection Algorithm 1 using a skeptic semantics, grounded or ideal, may result in a conclusive empty decision.*[5]

Proposition 2. *An agent calculating the goal-based action reflection Algorithm 1 using the credulous semantics: stable, may result in an inconclusive decision.*[6]

3.1 Support in Relation to the Zone of Proximal Development Using Formal Argumentation

In this section, based on the common-sense reasoning of activities using argumentation theory, we propose a theory to calculate the following four scenarios in assistive agent-based technology:

I. ZPD_i independent activity execution. This scenario describes an *observer agent* which takes the decision to do nothing to support a person. More formally, the type of fragments (Definition 3) generated by the agents are of the form $HF = \langle S, O', h^*, g \rangle$ such that $h^* \in \mathcal{H}_A = \{\emptyset, do\_Nothing\}$. In this setting, all the extensions generated by $SEM(AF_P) = \mathcal{E}$ during a period of time will create an activity structure. In other words, the cumulative effect of generating fragments, re-construct an activity in a bottom-up manner. Moreover, Algorithm 1 returns only values of α, *i.e.* the current value of a qualifier when the agent does not take any supportive action. This context defines the baseline of activity execution independence of a person.

II. ZPD_h activities supported by another person. Similarly to previous scenario, the role of the software agent is to be an observer. However, built fragments have the form $HF = \langle S, O^*, h^*, g \rangle$ such that $h^* \in \mathcal{H}_A = \{\emptyset, do\_Nothing\}$ and $O^* = O' \cup O''$, where O^* is the set of joint observations from the agent's perspective about the individual supported (O') and the supporter O''. We have that $O' \subseteq O''$, and $O', O'' \neq \emptyset$. In this scenario, O'' is considered a reference set of observations (**Ref** lines 3 and 12 in Algorithm 1). Algorithm 1 will return a value of α which measures to what extent an individual follows the guide provided by another person.

---

[4] Due to lack of space, the full proofs of these propositions are omitted.

[5] Proof sketch: output of grounded and ideal may include $\{\emptyset\}$. See [10].

[6] Proof sketch: output of stable semantics may include $\emptyset$. See [10].

When multiple extensions are collected during the period of time that the individual is supported, then a different set of activities than individual activity execution may be re-generated in a bottom-up manner.

III. ZPD_s activities supported by an agent. In this scenario, an *assistive agent* takes a decision oriented to uphold human interests, priorities and ability to conduct an activity. This is a straightforward scenario where $h \in \mathcal{H}_A \neq \{\emptyset, do\_Nothing\}$.

IV. ZPD_{h+s} caregiver and agent supporting cooperatively. In this scenario, the main challenge for the agent perspective is to detect: (1) actions that an assistant person executes, and (2) observations of both, the person assisted and the person who attends. This is similar to ZPD_H but with fragments built from $\mathcal{H}_A \neq \{\emptyset, do\_Nothing\}$. In this case, the level of ZPD_{H+S} is given by Algorithm 1, and the set of extensions $\mathcal{E}$ with *aligned* goals between agent and the caregiver.

4   Principles for Providing Consistent Assistive Services

In this section, we propose a set of general principles that AT based on deductive systems should follow to warranty consistency in their outputs.

4.1   Activity-Oriented Principles

Based on previous detailed analysis of different ZPD scenarios, we propose in the following a set of principles that need to be fulfilled to provide assistive services.

Proposition 3. *Let $\mathcal{A}^*$ be the set of all the possible activity models; let $R \subseteq \mathcal{A}_j$ a set of fragments from an activity model; let $h_j \in \mathcal{H}_a$ and $g_j \in \mathcal{G}_a$ be an agents' action and goal; and let $E_k \subseteq \mathcal{E}$ be an extension of hypothetical fragments. The following holds:*

$$\nexists \langle R, h, g \rangle in E_k \notin \mathcal{A}^*$$

Proposition 3 establishes that there is no hypothetical fragment that can be built that does not belong to the set of all the activity models. This proposition defines a principle of *closure*, *i.e.* that an AT system should not generate outputs (e.g. AT services) that are not contained in the main set of activities.

Proposition 4. *Let $\mathcal{A}^*$ be the set of all the possible activity models; and let $\mathcal{A} \langle Ax, Go, Op \rangle$ be an activity model with $\mathcal{A} \subseteq \mathcal{A}^*$. The following holds: $\nexists$ any $ax \in Ax$, $g \in Go$ or $o \in Op \notin \mathcal{A}^*$.*

Proposition 4 seems straightforward but it establishes that only those activities framed on an activity model can be seen as actions, goals or operations. Out of an activity, individually, those elements have not influence in the decision-making of an argument-based assistive system. Proposition 4 has a social science background, activity theory defines an activity by its motive, and activity necessarily builds on the hierarchy of actions and operations, roughly saying that there not exists any action, goal or operation out of an activity. These elements of an activity can not be considered separately or independently [21]. In this sense, Proposition 4 establishes the same principle, defining with Proposition 3 basic conditions of activity knowledge closure.

Postulate 1. *Let $\mathcal{O}_{Go}$ be a set of observations about human goals (Go) and actions (Ax) framed on an activity, captured by an agent using an activity model $\mathcal{A}$. Let $\mathcal{G}$ and $\mathcal{H}_A$ be agent's goals and its hypothetical actions. In order to provide non-conflicting assistance two properties have to be fulfilled:*

- *PROP1: $\mathcal{O}_{Go} \cap \mathcal{G} \neq \emptyset$*
- *PROP2: $\mathcal{O}_{Ax} \cap \mathcal{H}_A \neq \emptyset$.*

Postulate 1 can be seen as a *self-evident* rule that any intelligent assistive system should follow. PROP1 and PROP2 provides coherence among human-agents actions and goals. This two properties may define a first attempt to establish consistency principles of agent-based assistance. This is a future work in our research.

5 Implementation

The scenario selected for implementing a demonstrator for the formal results describes the situation where an older adult performs the the activity to *distribute medication* into a *medicine cabinet*. This activity is supported by an

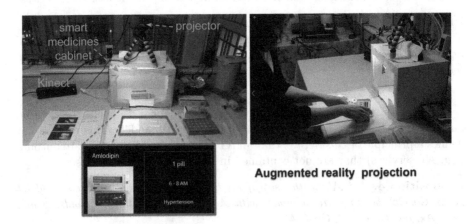

Fig. 4. Smart medicines cabinet using argument-based reasoning and an augmented reality projection. (I) Gesture recognition using three Kinect cameras, one for client body capture, another for assistant personal gesture recognition, last one (Kinect sensor 2) on the top of the cabinet to recognize text from medicines boxes; (II) Google API for text recognition; (III) argument-based reasoning; (IV) goal-based action reflection to consider human side; (V) database containing doses and timing of pill intake.

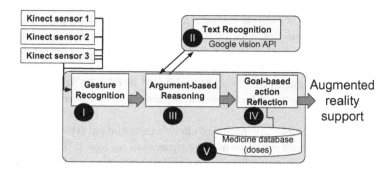

Fig. 5. Text

intelligent system and technology for augmented reality that is used for mediating the information provided by the system (see Fig. 4).

The prototype architecture consists of five main parts (see Fig. 5): (1) gestures recognition: obtaining observations from individuals using Kinect cameras; (2) text recognition using another Kinect camera with Google API text recognition (https://cloud.google.com/vision); (3) argument-based reasoning: the main agent-based mechanism of common sense reasoning; (4) goal-based action reflection generating an augmented reality feedback: a module to generate support indications as projections in the smart environment; and (5) a database of medicine doses to obtain appropriate messages.

We use three 3D cameras to capture: (1) observations of an individual that needs help in a physical activity; (2) observations of the smart environment, including a supporting person; and (3) information of the handle gestures of medicine manipulation. A central computer was connected to the cameras, processing the information in real-time analyzing gestures of individuals as observations for the agent. The agent platform (JaCaMo) was used to build the agent. An argumentation process was used using an argumentation library previously developed (see [16]). An agent updates/triggers its plan every time that a pre-defined gesture of the 3D camera is identified. Those pre-defined gestures were defined and trained based on data from three older adults and two medical experts.

6   Discussion and Conclusions

Our main contribution in this paper is a formal understanding of the interplay among an assistive agent-based software, a person to be assisted and a caregiver.

Argumentation-based systems, have become influential in artificial intelligence particularly in multi-agent systems design (see [8] for a systematic review). Argumentation theory can be seen as a process to provide *common-sense* to the decision-making process of a deductive system. Common-sense reasoning about an activity implies a non-monotonic process in which the output may change, when more knowledge is added. In the context of this paper, the contrary of a non-monotonic behavior is, for example a *stubborn system*, providing support when an individual does not need it and even under direct negative response from a user. In this paper we argue that non-monotonic reasoning may be used as main mechanism for decision-making of intelligent assistive systems. In fact, in ambient-assistive literature few authors have explore this approach (see [18, 25, 27, 28, 30]).

We propose an algorithm to integrate client's information (the activity model Definition 1) into the final decision-making process of an agent. This mechanism captured in Algorithm 1, resembles a process of "reflection" which in humans is a re-consideration of actions and goals given some parameters. Our reflection mechanism can be seen as an "action-filtering" process with the human-in-the-loop[7]. We also analyze different outputs of Algorithm 1 considering two groups of argumentation semantics (Propositions 1 and 2).

We propose different properties that software agents should follow if their goals are linked to human goals. We highlight the relevance of Postulate 1 which is understood as a primary rule for an intelligent assistive system. The relevance and impact of these properties not only covers agents based on formal argumentation theory, but other approaches, such as those based on the Belief Desire Intention model [5].

Our proposed principles are a starting point for evaluating assistive technology systems. This is a first step to establish general properties that such system should follow. We are aware that several principles can be added and we are aiming to continue this research line as future work.

We are also interested in the analysis of activity dynamics extending our formal results. In activity theory, the hierarchical structure is dynamic, there are transformations among internal levels of the hierarchy triggered by the demands and prerequisites in the environment [23]. We aim to investigate transformations in the activity, for example when the ZPD "increases", *i.e.* a person can achieve more activities with help of a caregiver or an assistive technology system, the activity hierarchy changes. From computational point of view, such change implies a modification at the information structure level, which may define scenarios where consistency can not be assured. In this sense, part of the future work will be focused on analyzing activity dynamics, but leveraged by the current "static" research of activities *e.g.* in [15, 17, 18, 29].

References

1. Aljaafreh, A.L., Lantolf, J.P.: Negative feedback as regulation and second language learning in the zone of proximal development. Mod. Lang. J. **78**(4), 465–483 (1994)
2. Amgoud, L.: Postulates for logic-based argumentation systems. Int. J. Approx. Reason. **55**(9), 2028–2048 (2014)
3. Bench-Capon, T., Dunne, P.E.: Argumentation in artificial intelligence. Artif. Intell. **171**(10), 619–641 (2007). http://www.sciencedirect.com/science/article/pii/S0004370207000793

---

[7] A concept to integrate human information in *cyber-physical systems* [35].

4. Besnard, P., Hunter, A.: A logic-based theory of deductive arguments. Artif. Intell. **128**(1–2), 203–235 (2001)
5. Bratman, M.: Intention, Plans, and Practical Reason. Harvard University Press, Cambridge (1987)
6. Bratman, M.E., Israel, D.J., Pollack, M.E.: Plans and resource-bounded practical reasoning. Comput. Intell. **4**(3), 349–355 (1988)
7. Caminada, M., Amgoud, L.: On the evaluation of argumentation formalisms. Artif. Intell. **171**, 286–310 (2007)
8. Carrera, Á., Iglesias, C.A.: A systematic review of argumentation techniques for multi-agent systems research. Artif. Intell. Rev. **44**(4), 509–535 (2015)
9. Chaiklin, S.: The zone of proximal development in vygotsky's analysis of learning and instruction. Vygotsky's educational theory in cultural context **1**, 39–64 (2003)
10. Dix, J.: A classification theory of semantics of normal logic programs: II. Weak properties. Fundam. Inform. **22**(3), 257–288 (1995)
11. Dung, P.M.: On the acceptability of arguments and its fundamental role in non-monotonic reasoning, logic programming and n-person games. Artif. Intell. **77**(2), 321–357 (1995)
12. Dung, P.M., Thang, P.M.: Closure and consistency in logic-associated argumentation. J. Artif. Intell. Res. **49**, 79–109 (2014)
13. Engeström, Y.: Learning by Expanding: An Activity-Theoretical Approach to Developmental Research, vol. 53. Orienta-Konsultit Oy, Helsinki (2015). https://doi.org/10.1017/CBO9781107415324.004
14. Guerrero, E., Lindgren, H.: Practical reasoning about complex activities. In: Demazeau, Y., Davidsson, P., Bajo, J., Vale, Z. (eds.) PAAMS 2017. LNCS (LNAI), vol. 10349, pp. 82–94. Springer, Cham (2017). https://doi.org/10.1007/978-3-319-59930-4_7
15. Guerrero, E., Nieves, J.C., Lindgren, H.: ALI: an assisted living system for persons with mild cognitive impairment. In: 2013 IEEE 26th International Symposium on Computer-Based Medical Systems (CBMS), pp. 526–527. IEEE (2013)
16. Guerrero, E., Nieves, J.C., Lindgren, H.: Semantic-based construction of arguments: an answer set programming approach. Int. J. Approx. Reason. **64**, 54–74 (2015)
17. Guerrero, E., Nieves, J.C., Sandlund, M., Lindgren, H.: Activity qualifiers in an argumentation framework as instruments for agents when evaluating human activity. In: Demazeau, Y., Ito, T., Bajo, J., Escalona, M.J. (eds.) PAAMS 2016. LNCS (LNAI), vol. 9662, pp. 133–144. Springer, Cham (2016). https://doi.org/10.1007/978-3-319-39324-7_12
18. Guerrero, E., Nieves, J.C., Sandlund, M., Lindgren, H.: Activity qualifiers using an argument-based construction. Knowl. Inf. Syst. **54**(3), 633–658 (2018)
19. Hedegaard, M.: The zone of proximal development as basis for instruction. In: An Introduction to Vygotsky, pp. 183–207. Routledge, Abingdon (2002)
20. Horvitz, E.J., Cooper, G.F., Heckerman, D.E.: Reflection and action under scarce resources: theoretical principles and empirical study. In: IJCAI, pp. 1121–1127 (1989)
21. Kaptelinin, V., Nardi, B.A.: Acting with Technology: Activity Theory and Interaction Design. Acting with Technology. MIT Press, Cambridge (2006)
22. Leontyev, A.N.: Activity and Consciousness. Personality, Moscow (1974)
23. Lindgren, H.: Activity-theoretical model as a tool for clinical decision-support development. In: 15th International Conference on Knowledge Engineering and Knowledge Management, Poster and Demo Proceedings in EKAW, vol. 215, pp. 23–25 (2006)

24. Lindgren, H., Baskar, J., Guerrero, E., Nieves, J.C., Nilsson, I., Yan, C.: Computer-supported assessment for tailoring assistive technology. In: Proceedings of the 6th International Conference on Digital Health Conference, pp. 1–10. ACM (2016)

25. Marcais, J., Spanoudakis, N., Moraitis, P.: Using argumentation for ambient assisted living. In: Iliadis, L., Maglogiannis, I., Papadopoulos, H. (eds.) AIAI/EANN -2011. IAICT, vol. 364, pp. 410–419. Springer, Heidelberg (2011). https://doi.org/10.1007/978-3-642-23960-1_48

26. Maslov, S.: Theory of Deductive Systems and Its Applications. MIT Press Series in the Foundations of Computing. MIT Press, Cambridge (1987)

27. Muñoz, A., Augusto, J.C., Villa, A., Botía, J.A.: Design and evaluation of an ambient assisted living system based on an argumentative multi-agent system. Pers. Ubiquitous Comput. 15(4), 377–387 (2011)

28. Muñoz, A., Serrano, E., Villa, A., Valdés, M., Botía, J.A.: An approach for representing sensor data to validate alerts in ambient assisted living. Sensors 12(5), 6282–6306 (2012)

29. Nieves, J.C., Guerrero, E., Lindgren, H.: Reasoning about human activities: an argumentative approach. In: 12th Scandinavian Conference on Artificial Intelligence (SCAI 2013) (2013)

30. Oguego, C.L., Augusto, J.C., Muñoz, A., Springett, M.: Using argumentation to manage users' preferences. Future Gener. Comput. Syst. 81, 235–243 (2018)

31. Pollack, M.E.: Intelligent technology for an aging population: the use of AI to assist elders with cognitive impairment. AI Mag. 26(2), 9 (2005)

32. Pollock, J.L.: Defeasible reasoning. Cogn. Sci. 11(4), 481–518 (1987). https://doi.org/10.1207/s15516709cog1104_4, http://doi.wiley.com/10.1207/s155 16709cog1104_4

33. Rao, A.S., Georgeff, M.P., et al.: BDI agents: from theory to practice. In: ICMAS, vol. 95, pp. 312–319 (1995)

34. Salomon, G., Globerson, T., Guterman, E.: The computer as a zone of proximal development: internalizing reading-related metacognitions from a reading partner. J. Educ. Psychol. 81(4), 620 (1989)

35. Schirner, G., Erdogmus, D., Chowdhury, K., Padir, T.: The future of human-in-the-loop cyber-physical systems. Computer 46(1), 36–45 (2013)

36. Schut, M., Wooldridge, M.: Principles of intention reconsideration. In: Proceedings of the Fifth International Conference on Autonomous Agents, pp. 340–347. ACM (2001)

37. Schut, M., Wooldridge, M., Parsons, S.: The theory and practice of intention reconsideration. J. Exp. Theor. Artif. Intell. 16(4), 261–293 (2004)

38. Vygotsky, L.S.: Mind in Society: The Development of Higher Psychological Processes. Harvard University Press, Boston (1980)

39. de Witte, L., Steel, E., Gupta, S., Ramos, V.D., Roentgen, U.: Assistive technology provision: towards an international framework for assuring availability and accessibility of affordable high-quality assistive technology. In: Disability and Rehabilitation: Assistive Technology, pp. 1–6 (2018)

Microsoft Hololens - A mHealth Solution for Medication Adherence

Martin Ingeson[1] , Madeleine Blusi[2] , and Juan Carlos Nieves[1(✉)]

[1] Department of Computing Science, Umeå University, 901 87 Umeå, Sweden
martiningeson@gmail.com, jcnieves@cs.umu.se
[2] Department of Community Medicine and Rehabilitation, Umeå University,
901 87 Umeå, Sweden
madeleine.blusi@umu.se

Abstract. The aim of this paper is to introduce a smart mHealth application based on the augmented reality (AR)-paradigm that can support patients with common problems, related to management of their medication. This smart mHealth application is designed and implemented as a medication coach intelligent agent, called *Medication Coach Intelligent Agent (MCIA)*. The *MCIA* has to manage different types of information such as the *medication plan (medication regime)* of the patients, medication restrictions, as well as the patient's preferences and sensor input data from an AR-headset. Considering all this information, the *MCIA* leads with holistic decisions in order to offer personalized and unobtrusive interventions, in an autonomous way, to the patients. From a long-term perspective, the *MCIA* should also evaluate its performance over time and adapt in order to improve its interventions with the patients. To show the feasibility of our approach, a proof-of-concept prototype was implemented and evaluated. In this proof-of-concept prototype, the *MCIA* has been embodied as a smart augmented reality (AR)-mHealth application in the settings of a Microsoft HoloLens. The results show a high potential for using the *MCIA* in real settings.

1 Introduction

Medication adherence is a global problem [12], which can be defined as the "extent to which a patient acts in accordance with the prescribed interval, and dose of a dosing regimen" [13]. Lack of medication adherence leads to patients not achieving sufficient health outcomes [25], and about 25–50% of the patients do not follow their prescriptions correctly [22]. Non-adherence has been estimated to a cost of 100–289 billion dollars a year for the U.S healthcare system [29]. From a medical perspective it has long been recognized that poor adherence to medical treatment is a substantial roadblock to achieving better outcome for patients [16]. Nonadherence to medication regimens affects both quality of life and length of life [15].

© Springer Nature Switzerland AG 2019
F. Koch et al. (Eds.): AIH 2018, LNAI 11326, pp. 99–115, 2019.
https://doi.org/10.1007/978-3-030-12738-1_8

Several attempts have been made in order to address this problem such as different mHealth-applications[1] and different types of robots. Even though some interesting robots are on its way such as *Pillo*[2], robots have by nature, the limitation of being more or less fixed in its location.

Some of the critiques towards current mHealth- applications is that they lack several basic adherence attributes [19], as well as persuasive techniques to engage people in the digital management of their disease [18].

Non-adherence can be either intentional or unintentional [11]. Unintentional non-adherence could, at least in theory, easily be addressed by sending reminders to patients. This may work very well for mHealth-applications, but it is more difficult when using robots since the user may not be close enough to the robot at all times. Intentional non-adherence is more challenging since sending a reminder can be seen as a useless attempt since, it will most likely not change the mental state of the patient. There is a conflict of interest between the system/agent and the patient in this situation. Persuasive techniques may play an important role when dealing with intentional non-adherence, but as mentioned many mHealth-applications lack this feature. AR[3]-headsets open up for a seamless interaction and brings a new approach for dealing with the problems of non-adherence.

An augmented reality (AR)-headset makes it possible to have the mobility of a mobile device and still establish a *social* and *friendly* relationship with the user. Through holograms it is possible to augment the users' field of view with an avatar. With a digital avatar there are more possibilities regarding the looks and appearance compared with a physical robot, which might have more affect on the intentions of the users. An AR-headset, such as the *Microsoft HoloLens*, also makes it possible to be more aware of the environment and user activities because of its many sensors.

Against this background, this paper introduces a novel solution to lead with the medication adherence problem based on the AR-paradigm and intelligent coaching systems. In particular, we introduce the so-called *Medication Coach Intelligent Agent (MCIA)*. The *MCIA* has *proactive* and *reactive* behavior in order to support the medical management of patients. Moreover, the *MCIA* has autonomous reasoning capabilities that allow the *MCIA* to lead with long-term goals in the settings of medication plans. As part of the results of this paper, an architecture of the *MCIA* is introduced. This architecture aims for a techno-logically scalable solution based on an AR-headset and multi-agent systems. We also present a usability evaluation of a proof-of-concept prototype of the *MCIA*.

The rest of the paper is organized as follows. In Sect. 2, different issues regarding medication management are discussed. In Sect. 3, the main goals of the research addressed in this paper from the medical perspective is presented. In Sect. 4, a theoretical framework regarding the *MCIA* is presented. In Sect. 5, an implementation of the *MCIA* in the settings of the *Microsoft HoloLens* is pre-

---

[1] Practice of medicine and public health using mobile devices, apps, smart devices and smartphones.

[2] https://www.pillohealth.com.

[3] Augmented reality.

sented. In Sect. 6, an evaluation of our proof-of-concept prototype is described. In Sect. 7, a short review of the related work is presented. In the last section, we outline our conclusions and future work.

2   Medication Scenario

The research in this project was developed as nurses from home healthcare brought attention to several patients having problems maintaining medication adherence through self management. Self management includes strategies and activities a person performs to live well with illness and it can be performed by the individual or in collaboration with a significant other [5,8,30]. Patients who are unable to perform health- and medication related activities as self management, for example handling and taking prescribed pills and following a medication plan, can get professional help in their homes, so called home healthcare [5]. A common reason for patients over age 70 to enroll in home healthcare is they are no longer able to handle their medication through self management and need professional help.

2.1   Patient Groups

From a medication management perspective, patients who use medicine regularly can be categorized into three conceptual groups. **Group 1** is independent and do not rely on help from others for managing their medication. **Group 2** is partly independent, receives help from relatives or friends, but do not get professional help. **Group 3** is in need of professional help.

The target group for the research in this project are patients from groups 2 and 3. The purpose is to investigate if AR-technology (using an AR-headset) may be used as a tool to increase their ability to improve and maintain medicine-related self management, thereby contributing to them staying independent for a longer time, delaying need for home healthcare and facilitate medication adherence.

2.2   Rules for Interchangeable Medicines

Many patients have several different medicines, as a strategy to simplify handling pills they use pill dispensers, where the medicine is distributed on a weekly basis. A common problem for the target group, and a reason why many patients need help managing their medication, is the continuous variation regarding their medicines, names of medicines and the visual appearance of pills and packages. This leads to patients and their non-professional helpers being confused when handling medication and preparing pill dispensers, which in turn leads to needing a nurse to come to their homes on a regular basis, preparing the dispenser for them.

What actually causes the variation is that the pharmacies can deliver different brands for the same type of medicine. The names and boxes varies with the brand

and this makes the patients insecure and afraid of preparing their dispensers. The underlying reason for the frequent exchange of medicine brands are the rules for interchangeable medicines which are applied in most European countries. According to the rules, if a patient gets prescription for a particular medicine, the pharmacy always has to offer the brand with the lowest price if the medicines are interchangeable [1].

One of the main aims of the *MCIA* is to make patients and their helpers confident enough, using an AR-headset, to prepare their dispensers and therefore remove or delay the need of home healthcare (nurse). Another priority of the *MCIA*, in order to consider a long term experience, will be to help the patients to follow their medication plans through self management.

2.3   Prescription and On Demand Medicines

For this project a characterization of medicines has been made since there are differences in how different types of medicines should be managed by the *MCIA*. All medicines comes with prescription from a doctor. Prescriptions include adherence information about how the medicine should be taken regarding dose and time schedule. Adherence information is personalized for each patient, printed on adhesive labels at the pharmacy and attached to each package of medicine.

Medicines to be Taken On a Regular Basis: The prescription label on these medicines state dose and the specific times each dose of the medicine should be taken. The goal of the *MCIA* is to make sure that the patient takes these medicines at the times they are specified.

Medicines to be Taken On Demand: This type of medicines are medicines that the patient can take when he or she feels the need. Examples of common on demand medicines are pills to decrease pain or anxiety. The information on the prescription label states strength per dose, minimum time interval between doses and maximum amount of doses allowed in 24 h. The goal of the *MCIA* will be to make sure that the user does not exceed the maximum dosage per day and occasion.

3   Research Questions and Methodology

The research conducted in this project was a collaboration between the Department of Community Medicine and Rehabilitation and Department of Computing Science at Umeå University in Sweden. The collaboration originated from clinical problems and challenges experienced in home health care environments.

The purpose of the project was to develop, test, implement and evaluate an intelligent system that can support patients in

1. adherence to prescribed medication,
2. assist the patient in filling the pillbox, and
3. provide patients with individual assurance and confirmation that they are taking the right pills at the right time.

The overall research question was: Would it be possible to use a medication coach intelligent agent to solve this? From a medical perspective the following five areas had been identified as essential: (1) patients maintaining self management of pills, (2) confirmation and patient security, (3) personalization, (4) mobility and (5) possible replacement of nurses.

3.1 Maintaining Self Management

When a person is on a medication regimen it is the individuals responsibility to follow the medication as prescribed. There is a number of tools available to assist and facilitate self management. Using a pillbox is one of the most common strategies for patients on treatments involving multiple pills to be taken at different times throughout the day. Instead of taking pills from multiple packages every time, the pillbox is filled up for a week at the time. A pillbox commonly has 28 compartments, distributed as seven horizontal compartments labeled Monday to Sunday, with 4 vertical slots for each day. The vertical compartments have labels suitable for the most common times to take pills during a day, such as morning, noon, evening and bed. As previously mentioned in Sect. 2.1 it may be confusing for patients to fill their pillboxes, as due to the rules of interchangeable medicines one pill in the middle of a week can change both name and appearance. Thus patients apply for home health care and nurse can fill the pill box instead. With this patients lose their independence.

3.2 Confirmation and Patient Security

Currently patients turn to nurses for help when they feel insecure about their medicines. A desire for the system was that it should be able to communicate with the patients regarding questions about their pills. Example of questions from patients: Is this the right pill? When shall I take this pill? How many shall I take? This feature would be useful as support when patients fill their weekly pillboxes. Another desire was that the system should be able to remind patients at the time pills are to be taken and give information about how many to take on each occasion.

3.3 Personalization

Each patient has their own individual prescribed medication plan. The system must be able to personalize the answers to the questions to each patients' medication list as well as individual needs for support. A medication plan can change over time and also temporary changes can occur. The system needs to be able to cope with such changes.

3.4 Mobility

Requirements for mobility were that patients should not be tied to a specific time and place when using the system. Further, patients should be able to decide when and where to use it and also bring it along when leaving the home.

3.5   Replacement for Nurses

In Sweden, as in several other countries, there is a lack of nurses. With many nurse-positions vacant the work environment become stressful. If technology could assist patients in filling their pillboxes, and handling related issues as described above, nurses would be able to re-arrange their time to work tasks where it is not suitable to replace nurses with technology. The demographic development where the proportion as well as the number of older people with care needs has rapidly increased, and will continue to do so, brings an increased demand for long-term care services. This will be a challenge as current supply is considered to be insufficient and inadequate in terms of meeting current and especially future needs for long-term care. The demographic development is one reason why it is important to develop technology in the health care sector [17].

4   Theoretical Framework

The aim of this section is to formally introduce both data sources and a multi-criteria decision making approach for supporting the decision making processes of the *MCIA*.

4.1   Data Modeling

Let us start introducing the basic definition of a time point. A time point is a time stamp $\langle Date, time\_clock \rangle$. $\mathcal{T}$ denotes all the possible time points. Now let us introduce the basic definition of a medicine. A particularly interesting attribute of a medicine is its Anatomical Therapeutic Chemical (ATC)-code. The ATC-classification is an internationally accepted classification system, based on active ingredients and their therapeutic, pharmacological and chemical properties[4]. By *ATC_codes*, we denote a finite set of ATC-codes. Hence, a medicine is defined as follows:

Definition 1. *Medicine*
A medicine m is a tuple of the form $\langle \epsilon, \varsigma, p, \delta, \alpha \rangle$, such that $m \in ATC\_codes \times \mathbb{R} \times [0,1] \times I \times Active\_ingridients$, where $\varsigma \in \mathbb{R}$ denotes a substance concentration in milligrams, $p \in [0,1]$ denotes a priority degree, $\delta \in I$ denotes a time interval such that $I = \mathcal{T} \times \mathcal{T}$. $\mathcal{M}$ denotes the set of all possible medicines.

Medicines will be managed in terms of events. An *event* is something that happens and which should be acknowledged by the *MCIA*. For example, if a reminder is presented to a patient whereupon the patient takes the medicines, the *MCIA* should notice that event and not present any more reminders regarding the same medicines for the same occasion. *Events* could also be things like eating food, drinking milk or other things that might have an impact, or have an effect, on the users' medication. This project however, only considers *events* regarding taking of an oral medicine, and is defined as follows:

---

[4] http://www.hpra.ie/homepage/medicines/medicines-information/atc-codes.

Definition 2. *Event*
An event e is a pair $\langle m, t \rangle$, such that $e \in \mathcal{M} \times \mathcal{T}$. $\mathcal{E}$ denotes the set of all possible events.

Constraints regarding medications can appear in three different ways. Firstly, it can be a medicine incompatibility, i.e. some medicines should not be taken at the same time as others, since it may have an impact on the effect of one or both of the medicines [6]. Secondly, there are time constraints. For example, if a patient takes an on demand medicine (such as regular painkillers), then a certain amount of time has to elapse until he or she can take it again. Thirdly, there can be a maximum dosage, or amount, per day. Due to lack of space, the formal definition of these constraints are not presented in this paper. The formal definition of these constrains can be found in [20]. We assume that the set of all possible constraints are denoted by *Constraints*.

A *medication plan* is the general plan which the *MCIA* wants the patient to follow. It is the foundation of the goals of the *MCIA*, and it is adherence to the medication plan that will be the primary source of feedback on how well the *MCIA* performs. The *medication plan* consists of all medicines that are prescribed, and also a set of constraints which should be considered while taking these medicines.

Definition 3. *Medication plan*
A medication plan MP is of the form $MP = \langle (m_1, m_2, \ldots, m_n), \mu \rangle$ such that $m_i \in \mathcal{M}(1 \leq i \leq n)$, $\mu = \{C_1, C_2, \ldots, C_n\}$, $C_i \in Constraints(1 \leq i \leq n)$.

Medication adherence is the measure of how well patients follow their medication plan. This is important information for the *MCIA*, since it can be seen as the result of its actions and decisions. *Medication adherence* can be divided into two parts, overall adherence (how well the patient is following the medication plan), and the individual adherence for a specific medicine. It is not easy to measure adherence since there are many factors which it depends on, e.g. skipping one medicine one time might be fine, while skipping another is not. Elementary factors of estimating the adherence are the *priority* of each medicine, which indicates how important it is to take the medicine, and the history of the intakes (compliance to the plan). The two definitions of adherence are presented below. An *adherence function* will be used to calculate adherence.

Definition 4. *Medication adherence for individual medicines*
Let β_m be the medication adherence of a medicine m such that $\beta_m = f_1(\gamma_m, p)$ where f_1 is an adherence function for individual medicines, γ_m is the history for medicine m and p be its priority.

Definition 5. *Medication adherence in general*
Let π be the over all medication adherence $\pi = f_2(\sum_1^n \beta_{m_i})$ where, f_2 is an overall adherence function, and β_{m_i} is the individual adherence for medicine $m_i(1 \leq i \leq n)$.

Let us point out that the adherence functions f_1 (w.r.t. Definition 4) and f_2 (w.r.t Definition 5) are basically distance functions between the current state of adherence and the intended medication plan. Hence, these functions can be implemented in different ways. In our proof-of-concept prototype, f_1 was implemented as a model checking function, based on weak-constraints, following Answer Set Programming (ASP) [7], regarding the constraints of each medication in the medication plan. Regarding f_2, it was implemented as a basic normalization function, $f_2 : \mathbb{R} \to [0,1]$, such that 0 means null adherence and 1 means total adherence.

4.2   Decision Making Modeling

In order to make decisions considering all of the relevant information such as the *medication plan* and the so called *information variables*, a *multi-criteria decision making-approach* has been chosen and more specifically, the *weighted sum method (WSM)* [21]. *Information variables* are used by the *WSM* for calculating weighted sums; an *information variable* is a pair of the form $\langle n, v \rangle$, where n is a propositional atom that describes what the variable represents (such as a preference or a context factor) and $v \in [0,1]$, e.g. $\langle prefersAudioOutput, 0.4 \rangle$, $\langle noisy, 0.8 \rangle$.

Information Variables are Compensatory. Having a global rank where a good criterion can compensate for a bad criterion is usually referred to as the *full aggregation approach* [21]. This is highly desirable since the *MCIA* will deal with conflicting information and priorities. For example, if the user is in a very noisy environment the *MCIA* should presents visual information, even if the user preference of visual presentation is very low. High values should therefore be able to compensate for low values, in order to make context-aware decisions.

Information Availability. One of the drawbacks with *multi-criterion decision making* in general, is that a lot of information has to be specified. In this case however, the information should always be available in real time through sensors and internal values.

All desires (also called goals), which the *MCIA* have committed to achieve, are called *intentions* and for each *intention* there is a finite set of *actions* $\{a_1, a_2, \ldots, a_n\}$. *Actions* are basically different ways of achieving an *intention*. *Actions* can also be seen as the *MCIA's* means of interacting with the environment. The distinction between *intentions* and *actions* is a way of handling high level reasoning (using *intentions*), while still being able to adapt and be sensitive to the current situation (by using an appropriate *action*). An *intention* is defined as follows:

Definition 6. *Intention*
An intention x is a pair $\langle ID, \alpha \rangle$, where $ID \in \mathbb{N}$, α be the intention to be performed.

Example 1. Let x_r be the *intention* to send a reminder to the user. Then there is a set of actions $\{a_1, a_2, a_3\}$ related the intention x_r, where $a_1 =$ use audio output, $a_2 =$ use visual output and $a_3 =$ use audio and visual output.

Before presenting the definition of decision making, a couple of related definitions are presented. *Utiliy weights* can be seen as the *priority* of the *information variable* regarding a given decision, and is defined in the following way.

Definition 7. *Utility weights*
Let w_{xd} be the weight for information variable x regarding the decision d, then $w_{xd} \in \{L, M, H\}$ (low, medium or high importance).

An exact numeric value of the *utility weight* for the different priorities, is not defined and it may have to depend on the type of decision. However, a value between 0 and 1 will always be used for each of the different levels of importance.

The utility function $U(a)$, uses *information variables* with *utility weights* to calculate the utility of an alternative a and is defined in the following way.

Definition 8. *Utility function*
*Let a be an action, and $\sigma_1, \sigma_2, \ldots, \sigma_n$ be the positive **information variables**, then $U(a) = \sum_0^n x_i w_i$ where, x_i is the value of the **information variable** $\sigma_i (1 \leq i \leq n)$, w_i be the weight of the information variable σ_i.*

Only *positive information variables* are used in the calculation. *Positive,* simply means that if the *information variable* has a high value, it should increase the utility for the given alternative. This is chosen for simplicity of the calculation, but it puts some requirements on what *information variables* there must be in order for the utility function to be fair. Competing alternatives should always depend on similar *information variables*, which means that they should have the same importance and have a similar purpose. This problem could be addressed by setting a weight which corresponds to the exact value of the importance of the variable, but it is hard to exactly define the importance of an *information variable* for a given decision. Instead, *utility weights* are merely a rough estimation of how important an *information variable* is.

Information which is not defined explicitly as a value (such as information in the *medication plan*), but may still be important when calculating the utility of the alternative, will be converted into an *information variable* using a separate function. This function varies depending on the type of information, but the result will be a number between 0 and 1 and will therefore be treated as a regular *information variable*.

The decision of choosing the best *action* is taken in real time by using the following definition.

Definition 9. *Decision making*
*Let D be a decision and $a_1, a_2, \ldots, a_n$ be competing actions, then $D = max(U(a_1), U(a_2), \ldots, U(a_n))$ such that $U(a_i)$ is a **utility function** which calculates the utility of $a_i (1 \leq i \leq n)$.*

A plan is simply a list of *intentions* which, if nothing changes, will be executed by the *MCIA*.

Definition 10. *Plan*
Let δ be a plan $\delta = (\nu, \theta)$, $\nu = [x_1, x_2, \ldots, x_n]$ where each $x_i (1 \leq i \leq n)$ be an intention, $\theta = [l_1, l_2, \ldots, l_n]$ be a list of dependencies such that $l_i = (\pi, \beta)$, $\pi \in \{ID | (ID, \alpha)\ appears\_in\ \nu\}$, $\beta \subseteq \{ID | (ID, \alpha)\ appears\_in\ \nu\}^5$ and $\pi \notin \beta$.

5   Implementation

In this section, a proof-of-concept prototype of the *MCIA* is described. In this proof-of-concept prototype, the *MCIA* has been embodied as a smart augmented reality (AR)-mHealth application in the settings of a *Microsoft HoloLens*. This AR-mHealth application was designed as a *long-term experience application (LTEA)* [20]. The goal with the presented architecture and reasoning loop is to make the *MCIA* context-aware, unintrusive, being able to personalize to individual users and to plan towards long-term goals. These four areas are identified as particularly important when designing long-term experience applications for augmented reality [20].

The internal reasoning process of the *MCIA* follows the *beliefs-desires-intentions (BDI)*-model [31]. The *BDI* approach was chosen to handle a practical reasoning algorithm. Unity and Visual Studio were used to implement the prototype, and Vuforia was used as a plug-in to Unity in order to recognize medicine boxes. The general architecture of the system is depicted by Fig. 1. The architecture consists of three major components, the *MCIA*, *external agents* and *databases*. The *external agents* and the *databases* provides the *MCIA* with the information it needs in order to supply the services to the user. The external agents, which are also *BDI-agents*, were introduced in our previous work [27].

The reasoning loop of the *MCIA* can be seen in Fig. 2. A *plan of intentions* (later referred to as *plan*) will be constructed using the current state, referred to as *internal state* (Figs. 1 and 2), and the *long-term goals*. This *plan* will be created on a daily basis and planning will take place over a specific time period. By practical reasons, it is assumed that this planning process will take place during night time. This means that when the user wakes up, the *plan* for the day has already been made and only re-planning using the *event-driven approach* is necessary. The reason why it is referred to as an *event-driven process* is that *events* can be seen as triggers that changes the *internal mental state* of the *MCIA*. Therefore, in the case of an *event*, the *MCIA* should check for interactions with the *plan* and re-plan accordingly.

Proactive behavior emerges by *actions* executed in order to achieve the *intentions* of the *MCIA*, such as sending reminders to the patient. Reminders are also context-aware regarding time, but in our proof of concept, also by simulated information from the environment. *Reactive behavior* emerges by using input to directly trigger some behavior, e.g. using voice commands and information

---
^{5} *appears_in* is the classical membership operator in lists.

Application Architecture

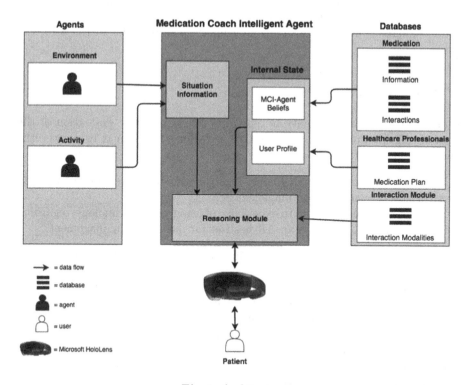

Fig. 1. Architecture

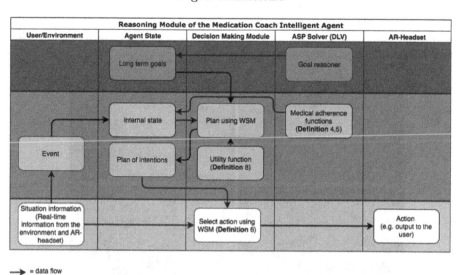

Fig. 2. Reasoning loop

regarding the user's vision (using the AR-headset), it is possible to display information regarding medicine boxes. *Autonomous behavior* emerges by reasoning about the *intentions of the day*. The purpose of *intentions* are to improve medication adherence of the user, and by evaluating the behavior of the user, the reasoning can be adapted.

6   Evaluation

In order to show the feasibility of our approach, a usability evaluation of the proof-of-concept prototype of the *MCIA* was done. We aimed to answer the following questions:

- Is there a difference, related to age, regarding if people are willing to use an AR-headset for medication management?
- Is there a difference, related to experience of using smart technology, regarding if people are willing to use an AR-headset for medication management?

The functionality involved displaying information about medicine boxes regarding two features namely, helping the user to use a medicine at this moment and to help the user to prepare a dispenser. The evaluation involved 15 participants who were selected by using the following criteria: a. Different levels of management of medication on a regular basis; b. Wide range of ages (medication management in applies to people in all ages, not just elderly); c. Different experiences using smart technology in general. The setting was a quiet and home-like environment. The participants were able to use voice, vision and gestures to interact with the system and were presented with both visual and audible output. Figure 3 summarizes the visual information presented to the participants. After the test they were asked to fill in a form. Responses were on a five-point *Likert-type scale* graded from 1 (strongly disagree) to 5 (strongly agree). The lower bound to agree was made at 4 (4 or 5 = agree).

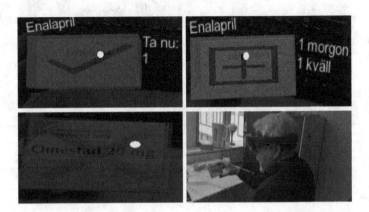

Fig. 3. Visual information about medicines presented to test participants

The evaluation showed that of all participants 20% perceived the technology hard to use and 13% thought that they would need a lot of training before using this technology in real life. There were a couple of vast differences regarding participants over and under 70 (4 and 11 respectively), and also between experienced smart technology users and those less experienced (8 and 7 respectively). Of the participants over 70, 50% were willing to use the technology in the future but none thought other people would appreciate the technology. For the participants under 70, the corresponding numbers were 91% and 100%. For participants over 70, 0% considered themselves as experienced smart technology users, while 73% of the people under 70 considered themselves as experienced. Of all experienced participants 100% were willing to use this technology in the future and 88% thought that most other people would appreciate the technology.

7   Related Work

The possibilities of using smart-glasses (AR) within a system to assist doctors and other healthcare-personnel in emergency situations was explored in [14]. Smart-glasses was connected to different types of medical equipment and was used to display important information for the person wearing them. The smart-glasses could also be used to record video/audio and to take snapshots of the process.

Mitrasinovic *et al.* concluded that smart-glasses have evident utility to healthcare professionals [26]. A major advantage mentioned by Mitrasinovic *et al.* is that the glasses are hands-free which liberates the users from giving manual input.

A concept to send context-aware reminders to users in order to increase medication adherence was presented in [24]. They argued that sending reminders should depend on other factors than time, since there are a lot of scenarios where time-based reminders can fail. Results showed that the concept proved to be better compared to time-based reminders, which motivates the need of being context-aware. Also, one of their priorities was to be unobtrusive, in this case in terms of monitoring of elderlys' activities, which we also believe is important. The user should not be annoyed by the system and it is important that the user has a positive attitude towards the system, especially for the results of long-term usage.

An article about using a humanoid robot to support elderly peoples' everyday life [28], supplied similar functionality that we wish to do. They wanted to help the user to take the right medicine (selecting), prevent the user from taking the wrong medicine, helping the user remembering to take the medicine, keep a record of all medicine intakes and allow administers to remotely edit prescriptions. Their prototype showed that there is potential in using a humanoid robot, and by using a robot (*NAO* in this case) it is possible to handle additional problem domains such as emergency situations. We believe that an AR-headset, such as the *Microsoft HoloLens*, is even more capable of handling other problem domains requiring mobility and portability.

General tools to increase medication adherence for patients are smart dispensers (e.g. [4]), robots (e.g. [3]) and applications to mobile devices (e.g. [2]). General functionality of these devices is to remind the patient to take the medicine and to help them with taking the right medicines. Our vision is to combine all of these common features and to add a more intelligent behavior.

8   Conclusions and Future Work

Medication adherence is a healthcare issue that affects both youth and elderly patients around the world. Until now, there is no a general solution that can support the dynamic demands that each individual requires for keeping his or her self-medication management optimal. In this regard, we argue that our approach based on intelligent coaching systems and AR-headsets shows a solid and scalable solution for leading with the complex processes of tailored services on medication management. Results from our evaluation showed that participants felt comfortable using an AR-headset during medication management procedures, such as taking pills and putting pills into pill dispensers. The evaluation indicates that the *MCIA* embodied in an AR-headset can be a useful tool in helping patients to maintain self management and medication adherence. From a societal perspective, maintained self management is likely to delay or possibly prevent, the need for professional assistance by nurses. As most western countries suffer from lack of nurses and other health care professionals the effects of the *MCIA* would potentially have a high impact on sustainability of public health resources.

The feasibility of using the *MCIA* today is somewhat restricted to the limitations regarding wearable AR-technology hardware. Some of the most crucial limitations being short battery life, size of field of view and the physical size of the headset.

When replacing human interaction with technology devices the risk for possible negative effects must be taken into consideration. From a health-care perspective there are both medical and social risks. Changing a medical situation where traditionally health professionals are physically present in the patient's home and perform a task for the patient, to the patient performing the task themselves requires a shift of mindset. Changing routines is often challenging and may be particularly difficult as the supportive technology is inside the Hololens and thereby may be experienced as "invisible" by patient, as opposed to a nurse standing in front of him. Patients need to learn to trust the technology and the technology must be safe. If the system is not sufficiently programmed the patient will get the wrong information, thereby not taking the prescribed medication correctly which can lead to serious medical conditions. Among elderly home care clients loneliness is a recognized health problem. For many of them the visits from health professionals are the only social interaction they have with other humans. Replacing human interaction with technology devices may increase feelings of loneliness for those patients [9].

From a software perspective, providing proactive and autonomous behavior in a context-aware and personalized manner is of course also a challenge. Goal reasoning and planning are hard and complex processes which are necessary for proactive and autonomous behavior. Computationally heavy processes combined with already insufficient batteries are not a good match. One workaround to this problem can be to take advantage of the increased availability of Wifi and increased speed of mobile networks to offload the heavy computations from the mobile devices to a server. This would decrease the performance demands on the AR-headsets, hypothetically making them smaller, and enables more advanced and heavy processes since the limitations now lies at the server.

A majority of the patients in the target group for this mHealth device is likely to be found among persons older than 65 years. Statistics show that individuals in this group have 3 chronic conditions on average and over 70% take 5 or more drugs every day [10,23]. In our future work, we aim for a complete implementation of the *MCIA* and a long term usability evaluation. The evaluation presented in this paper was conducted with test persons in a living lab, set up as a quiet and home-like environment. Future evaluations are planned to take place in collaboration with a regional home-care organisation, where home-care patients with daily multiple-drug use will test the mHealth device at their kitchen tables based on their own individual medication plans. These evaluations will address usability and user experience from both technical and medical points of interest. We believe it has been a strength to this research project to have a multidisciplinary research group where researchers with experience from various fields such as computer science, AI, engineering, medicine and nursing have collaborated.

Acknowledgements. This research has been supported by Nordic Telemedic Center (EU Interreg Botnia Atlantica), SFO-V Strategic Research Area Health Care Science and CEDAR (Centre for Demographic and Aging Research).

The authors are very grateful to the anonymous referees for their useful comments.

References

1. Apoteket—på recept. https://www.apoteket.se/kundservice/receptlakemedel-sa-fungerar-det/
2. Medisafe—mobile application. https://medisafe.com
3. Pillo Health. Pillo: Your Personal Home Health Robot. https://www.pillohealth.com
4. Posifon caruousel—medicin doserare. http://careousel.se
5. National Board of Health and Welfare: Next of kin as family caregivers – extent and consequences. Socialstyrelsen National Board of Health and Welfare, Stockholm (2012). (In Swedish: Anhöriga som ger omsorg till närstående - omfattning och konsekvenser, SOSFS 2009:6)
6. Rounds, I.V.: What you should know about drug compatibility. Nurs. Clin. Med. **38**, 15 (2008)
7. Baral, C.: Knowledge Representation, Reasoning and Declarative Problem Solving. Cambridge University Press, Cambridge (2003)
8. Battersby, M., Lawn, S., Pols, R.: Conceptualisation of Self-management. Wiley, Oxford (2010)

9. Blusi, M., Kristiansen, L., Jong, M.: Exploring the influence of internet-based caregiver support on experiences of isolation for older spouse caregivers in rural areas: a qualitative interview study. Int. J. Older People Nurs. **3**, 211–220 (2014)

10. Bodenheimer, T., Berry-Millett, R.: Follow the money - controlling expenditures by improving care for patients needing costly services. N. Engl. J. Med. **361**(16), 1521–1523 (2009)

11. Choi, A., Lovett, A.W., Kang, J., Lee, K., Choi, L.: Mobile applications to improve medication adherence: existing apps, quality of life and future directions. Adv. Pharmacol. Pharm. **3**(3), 64–74 (2015)

12. Costa, E., et al.: Interventional tools to improve medication adherence: review of literature. Patient Prefer. Adherence **9**, 1303–1314 (2015)

13. Cramer, J.A., et al.: Medication compliance and persistence: terminology and definitions. Value Health **11**(1), 44–47 (2008)

14. Croatti, A., Montagna, S., Ricci, A.: A personal medical digital assistant agent for supporting human operators in emergency scenarios. In: Montagna, S., Abreu, P.H., Giroux, S., Schumacher, M.I. (eds.) A2HC/AHEALTH -2017. LNCS (LNAI), vol. 10685, pp. 59–75. Springer, Cham (2017). https://doi.org/10.1007/978-3-319-70887-4_4

15. Cutler, D.M., et al.: The value of antihypertensive drugs: a perspective on medical innovation. Health Aff. **26**(1), 97–110 (2007)

16. Cutler, M., Everett, W.: Thinking outside the pillbox-medication adherence as a priority for health care reform. N. Engl. J. Med. **362**(17), 1553–1555 (2010)

17. European-Comission. The 2009 ageing report: economic and budgetary projections for the EU-27 member states (2008–2060). European Comission - DG for economic and social affairs, economic policy committee. Technical report, European Comission (2009)

18. Geuens, J., Swinnen, T.W., Westhovens, R., De Vlam, K., Geurts, L., Abeele, V.V.: A review of persuasive principles in mobile apps for chronic arthritis patients: opportunities for improvement. JMIR mHealth uHealth **4**(4) (2016). https://www.ncbi.nlm.nih.gov/pmc/articles/PMC5083846/

19. Heldenbrand, S., et al.: Assessment of medication adherence app features, functionality, and health literacy level and the creation of a searchable web-based adherence app resource for health care professionals and patients. J. Am. Pharm. Assoc. **56**(3), 293–302 (2016)

20. Ingeson, M.: Long-term experience applications for augmented reality - in a medication adherence scenario. Master's thesis, Department of Computing Science, Umeå University (2018)

21. Ishizaka, A., Nemery, P.: Multi Criteria Decision Analysis: Methods and Software, 2nd edn. Wiley, Hoboken (2013)

22. Iuga, A.O., McGuire, M.J.: Adherence and health care costs. Risk Manag. Healthc. Policy **7**, 35 (2014)

23. Koper, D., Kamenski, G., Flamm, M., Böhmdorfer, B., Sönnichsen, A.: Frequency of medication errors in primary care patients with polypharmacy. Fam. Pract. **30**(3), 313–319 (2012)

24. Lundell, J., et al.: Continuous activity monitoring and intelligent contextual prompting to improve medication adherence. In: 29th Annual International Conference of the IEEE Engineering in Medicine and Biology Society, EMBS 2007, pp. 6286–6289. IEEE (2007)

25. Miller, N.H., Hill, M., Kottke, T., Ockene, I.S.: The multilevel compliance challenge: recommendations for a call to action. Circulation **95**(4), 1085–1090 (1997)

26. Mitrasinovic, S., et al.: Clinical and surgical applications of smart glasses. Technol. Health Care **23**(4), 381–401 (2015)
27. Nieves, J.C., Lindgren, H.: Deliberative argumentation for service provision in smart environments. In: Bulling, N. (ed.) EUMAS 2014. LNCS (LNAI), vol. 8953, pp. 388–397. Springer, Cham (2015). https://doi.org/10.1007/978-3-319-17130-2_27
28. Sermeus, W., et al.: Robotic assistance in medication management: development and evaluation of a prototype. In: Nursing Informatics 2016: EHealth for All: Every Level Collaboration-From Project to Realization, vol. 225, p. 422 (2016)
29. Viswanathan, M., et al.: Interventions to improve adherence to self-administered medications for chronic diseases in the united states: a systematic review. Ann. Intern. Med. **157**(11), 785–795 (2012)
30. Wagner, E.H., Bennett, S.M., Austin, B.T., Greene, S.M., Schaefer, J.K., Vonkorff, M.: Finding common ground: patient-centeredness and evidence-based chronic illness care. J. Altern. Complement. Med. **11**(Suppl 1), s-7 (2005)
31. Wooldridge, M.: An Introduction to Multiagent Systems. Wiley, Hoboken (2009)

A Knowledge-Based Simulation Framework for Decision Support in Brazilian National Cancer Institute

Antônio Augusto Gonçalves[1,2(✉)] ⓘ,
Sandro Luís Freire de Castro Silva[1] ⓘ,
Carlos Henrique Fernandes Martins[1] ⓘ, Cezar Cheng[1] ⓘ,
and José Geraldo Pereira Barbosa[2] ⓘ

[1] Instituto Nacional do Câncer - COAE Tecnologia da Informação,
Rua do Resende 195, Rio de Janeiro 20230-092, Brazil
augusto@inca.gov.br
[2] Universidade Estácio de Sá - MADE,
Avenida Presidente Vargas 642, Rio de Janeiro 200071-001, Brazil

Abstract. Knowledge Management is decisive for clinical decision-making and for delivering better outcomes for patients care. The importance of medical knowledge has been emphasized in the researches to support evidence-based medicine. Currently, cancer is responsible for over 130,000 deaths every year in Brazil. Extensive waiting queues for diagnosis and treatments have become routine. One of the critical success factors in a cancer treatment is the early diagnosis. The reduction of waiting time to start cancer treatment is one of the main issues for improvement of patient's quality of life and possibilities of cure. This study presents a knowledge-based simulation framework developed at the Brazilian National Cancer Institute (INCA) to reduce patients' waiting time to start cancer treatment.

Keywords: Knowledge-based simulation · Decision support ·
Framework for decision support

1 Introduction

Nowadays, a great number of countries deal with grave problems and substantial expenditures in healthcare organizations, rising from the increase in the demand for healthcare services due to the growth in the number of elderly citizens with chronic diseases. The cancer treatment planning has become more complex with a huge demand for accessibility to hospital services with efficiency, equality, and customization of care [1].

Healthcare processes are characterized by knowledge-intensive tasks, and Information and Communication Technology (ICT) is largely used to support knowledge management in healthcare organizations. Knowledge is extremely important in healthcare organizations because low-quality information can lead to deficient clinical decisions and even endanger lives [2].

© Springer Nature Switzerland AG 2019
F. Koch et al. (Eds.): AIH 2018, LNAI 11326, pp. 116–126, 2019.
https://doi.org/10.1007/978-3-030-12738-1_9

In this circumstances, Knowledge Management is decisive for clinical decision-making and to offer better outcomes for patients. The importance of medical information has been emphasized in the researches to support evidence-based medicine. Gradually, healthcare staff depends on ICT tools to deal with the challenges to create, structure, share and apply knowledge [3, 4].

Taking into consideration the principal objective of Knowledge Management (KM) in clinical decision-making, it is fundamental to use ICT tools to support effective sharing of information among to the clinic staff. Accessibility to actualized and accurate information is crucial. The use of ICT supports the storage of a big volume of data, which can be processed by decision support systems in healthcare services [5].

Cancer is responsible for over 130,000 deaths per year in Brazil. Advances in life quality increased citizens' life expectations. However, because of limited resources, cancer in Brazil can be considered a severe public health problem. The management of cancer treatment is a long and complex process. The reduction of the patient's waiting time to start cancer treatment plays an increasingly important role in the treatment of this chronic illness [5, 6].

Healthcare simulation models generally require the implementation of systems with complex activities, involving stakeholders with a diversity of views and intentions [7]. It is thought that the active involvement of clinic staff throughout the study can decrease these problems, creating solid ownership of the model formulation and acceptance of charge for actions to be taken [8].

Organizations rely on data analytics to strategic decisions making. Descriptive analytics is commonly used to provide insight into past behavior. However, greater value can be achieved by predicting future behavior. Knowledge-based simulation represents an innovative area linking the fields of computer simulation and artificial intelligence. Simulation plays an important role in predictive analytics [9, 10].

The objective of this article is to describe the development of a knowledge-based simulation framework for decision support applied in the Brazilian National Cancer Institute (INCA) to reduce patients' waiting time to start cancer treatment. The adopted methodology was focused on the patient treatment flow and on the quick start of cancer treatment. The theory of constraints was used to identify bottlenecks in patient treatment flow and a discrete event simulation model was created to exploit the system's constraints and produce ongoing improvement efforts.

2 Methods

Knowledge Management is important for the clinical decision-making process and for delivering better outcomes for patients. The knowledge-based framework deployed at INCA includes four steps: creating, structuring, sharing and applying. The four stages support an integrated simulation environment for cancer treatment planning. Fig. 1 presents a list of activities and tools contained within each step. The graphics configuration of this simulation framework is presented in Fig. 1.

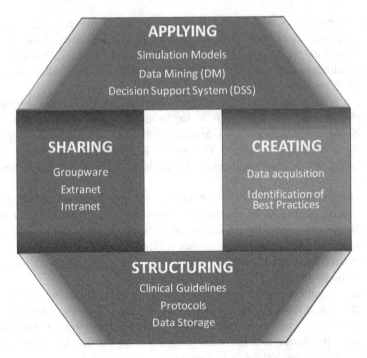

Fig. 1. INCA knowledge management process

The creating phase consists of data acquisition and identification of best practices. The process for data acquisition, from internal and external sources, is strongly interrelated with the clinical staff involved in cancer treatment.

The structuring phase consists of define, store, index and link clinical guidelines for cancer treatment. This phase involves classify and store protocols and clinical guidelines into data storages.

The sharing phase comprises the diffusion of best practices using extranets, intranet, groupware, communities of practice and multidisciplinary teams.

The last phase includes activities related to the clinical decision-making process and problem-solving using simulation models and Data Mining applications.

Computer simulation is an efficient tool to analyze complex systems and investigating different scenarios related to patient waiting-time reduction, resources allocation, staff scheduling, capacity planning, and on-going improvement.

A significant increase in the number of simulation models applied in healthcare services is evident. This growth is motivated by the ability of these simulation models in addressing complex problems that cannot be addressed by decision support systems. Much of its growth can be attributed to ICT development and the volume of real-time data available for analytics.

To support the physicians' activities, INCA's ICT technicians have developed data mining applications to find out the most effective treatments. These applications compare treatments patterns, symptoms and undesirable effects and then keep investigating which medical procedures will be most effective for a group of patients. This is also a way to identify the clinical best practices and protocols for cancer care.

Data mining have a high potential for cancer care institutions by allowing managers to consistently use data science and analytics tools to detect inefficiencies and apply good practices that increase outcomes and reduce costs. The INCA objective was to develop a framework to support the clinical decision-making process. Most of these tools are patient-centric as shown in the INCA Knowledge-based Simulation Framework presented in Fig. 2.

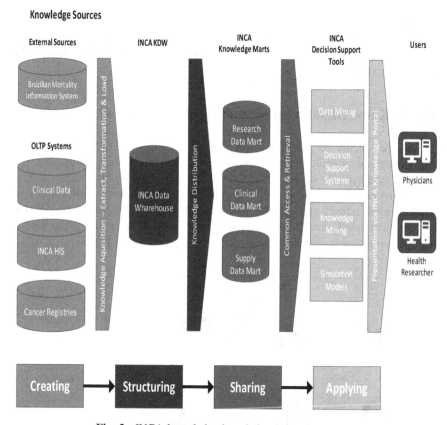

Fig. 2. INCA knowledge-based simulation framework

An innovative approach to analyzing issues of the healthcare services quality assessment is the creation of data-driven models. Such models allow considering the interests of all stakeholders to assess operations environment from the staff view and the effectiveness of work from the manager view. The critical success factor is to work with data from the hospital information system (HIS) and electronic health records (EHR).

This Knowledge-based simulation framework has developed using INCA's intranet network that provides safe access to key benefits of an e-health strategy that uses emerging information and communication technology to improve and facilitate health care. This solution has included enhanced collaboration between physicians and managers, simplifying the physician work, and empowering managers with sophisticated and cost-effective applications on the Web architecture. The legacy systems have fed the INCA data warehouse, which was the base to build the clinical data marts.

INCA data warehouse has processed information of 540,000 patients extracted from legacy systems. This repository was the hub of all data analysis. Several queries were created to validate data consistency. The development of a multidimensional database was to improve analytics applications and to examine the key performance indicators.

Over the last few years, the Brazilian National Cancer Institute (INCA) has been investing significantly in the implementation of an ICT architecture that integrates the organization's main processes and provides decision support systems which contain the following components.

3   INCA Knowledge-Based Simulation Framework

INCA has five specific hospital units with different stakeholders, but which share the same processes and technologies based on a common patient database and standardized information systems. To support the physicians' activities, several tools, such as tracking mechanisms for keeping the longitudinal patient history, online tools for gathering clinical information and the traditional medical record, are used. Most of these are patient-centric and make the hospital environment amenable to the kind of knowledge management system framework, such as presented in Fig. 3.

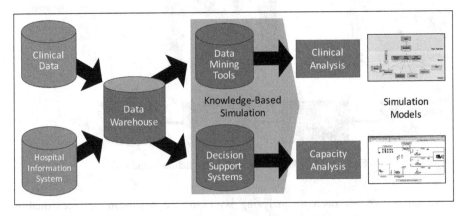

Fig. 3. INCA knowledge management system framework

The INCA knowledge-based simulation framework was developed using INCA's intranet network that provides a safe access to applications developed to improve healthcare. This solution includes a collaboration environment between physicians and managers, simplifying the physician work, and empowering managers involved in the decision-making processes. The legacy systems feed the clinical data repository, which is the basis to build the decision support systems.

3.1   Clinical Analysis (Patient Treatment Flow)

The proposed framework allows decision-makers to have a set of alternative scenarios in real-time. It consists of several models linked to the patient's treatment flow. The automated cancer treatment flow component is a new interesting feature. It offers the possibility to electronically generate the patient's treatment flow from the clinical data repository. Therefore, the users are able to examine, in a visual fashion, the evolution of the treatment. This component is a very useful tool to support decision-making with regards to the care provided. The doctors can blend, on one screen, the past, the present and the future events of the patient treatment history. The sequence of events, the dates and the duration of each event are very important to understand the structure of the treatment. Figure 4 shows an example of the flow of treatment for a particular patient.

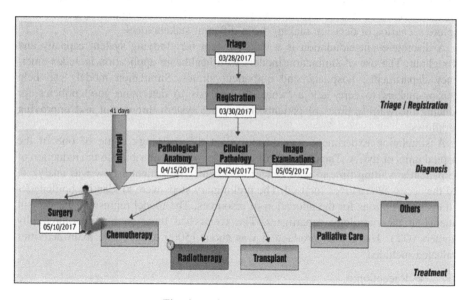

Fig. 4. Patient treatment flow

This component increases the traceability and is totally patient-oriented. It is possible to see, in an animated fashion, the details of the flow of a particular patient over the treatment process. The physician in charge of the case is able to follow a particular patient or a group of patients step by step in their cancer treatment. Based on the easily available information, one is able to detect and/or predict possible problem in the treatment flow.

Understanding the flow of the treatment, evaluating the constraints and managing the bottlenecks could be a possible way to improve quality. A constraint is anything that limits the system's performance. The identification of the constraints is a great opportunity to evaluate and improve the system concerned. There are five fundamental steps to follow: 1. Identify the system's constraint(s); 2. Decide how to exploit the system's constraint(s); 3. Subordinate/synchronize everything else to the above decisions; 4. Elevate the system's constraint(s); 5. If in the above steps the constraint has shifted, go back to step 1.

These concepts have already been explored for productivity improvements in the manufacturing area. The application of this approach to health services is suitable. A search in the literature provides an excellent and more detailed review of applying the theory of constraints (TOC) to healthcare services [11].

3.2   Capacity Analysis (Simulation Model)

Simulation modeling in healthcare is an efficient approach to analyze the interdependence between human resources and infrastructure variables in complex systems and to explore scenarios of decision-making from different stakeholders.

A discrete-event simulation is a valuable tool for studying system capacity and throughput. The use of simulation models with healthcare application includes emergency departments, hospitals, and outpatient clinics. Simulation models can help decision-makers to carry out a 'what if?' analysis to determine good policies for reducing the waiting times of patients, increasing system throughput and improving workflows.

A simulation experiment was developed at the radiology clinic of one of the hospital units of INCA. The objective of the model was to contribute to the reduction of the patient's waiting time to start cancer treatment. The patient's flow was analyzed, and the access alternatives focused. The preliminary steps were to identify bottlenecks and evaluate options for the allocation of resources. The model represents the flow of patients in the radiology department. The sector had three computed tomography scanners (CT). The annual production was about 15000 exams. The main activities evaluated included:

- Reception;
- Patient preparation;
- Medical examination;
- CT Scan;
- Film production.

The simulation model was used to examine alternative scenarios. The objective was the reduction of the waiting time between the computed tomography exam booking and its accomplishment. The target was to increase the capacity to make computed tomography (CT) exams. Figure 5 shows the model of the computed tomography exams' flow in the radiology clinic.

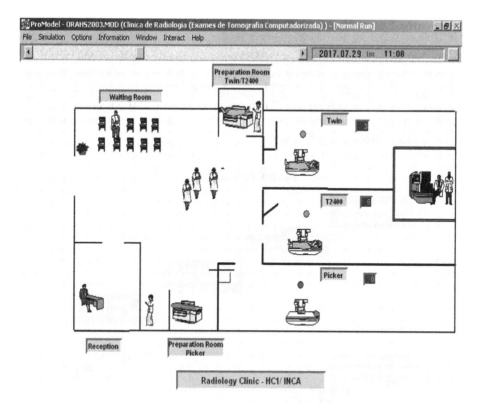

Fig. 5. Simulation model

This simulation model has supported decision makers through "what-if" scenarios for decision-making under conditions of high risks, uncertainty, and lack of information. Different scenarios were chosen for continually evaluating and improving upon key processes. They were characterized by different external environmental situations like a regular workload and an emergency workload at the radiology clinic. The first scenario was regular; it means that there were no unexpected additional patients in the radiology clinic. This scenario has provided management decisions about staff schedules and other resources like the numbers of computed tomography equipment. The emergency scenario has supported decision making in unpredictable circumstances in case of additional patients or equipment breakdown.

The method of conducting the experiment is the visual simulation that is proved to be a powerful tool. The proposed model examines individual patients as they arrive and pass through CT exams. Arrivals, priority rules and exam types provide the necessary detail to reproduce the real-life processes. Alternative scenarios can be compared and modified without high costs. The knowledge gained from the experiments allows one to take decisions without investing major resources.

4   Results

The Discrete Event Simulation model (DES) was used to investigate several "what- if" scenarios. The experiment showed that to reduce overall exam execution time was necessary to remove the phase of film production. The recommendation was to implement a Picture Archiving and Communication system (PACS). Figure 6 shows how the time of accomplishment of a group of examinations was reduced with the exclusion of the film production phase.

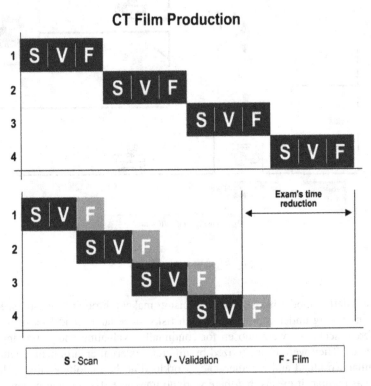

Fig. 6. Reduction of computed tomography (CT) exams time

All patients' data were aggregated in a clinical Data Mart to create the automated patient's treatment flow. This flow has joined electronic medical records and the sequence of events of a patient in only one screen. This approach was radically innovative allowing physicians to examine the clinical evolution of a patient quickly by using past events, current situation, and future procedures.

The dashboard has increased the traceability and was totally patient-oriented. It was possible to see, in an animated fashion, the details of the flow of a particular patient over the treatment process. The doctor in charge of the case was able to follow a specific patient or a group of patients step by step in their cancer treatment. Based on the easily available information, one was able to detect and/or predict possible bottlenecks in the treatment flow.

These analytics applications were developed to support clinicians' access robust data visualizations on their own, enabling them to *drill down* into and filter data based on their specific information needs.

The use of simulation model recommendations has made a significant impact in terms of reduction of the waiting time to obtain CT image diagnosis at INCA. A comparative study has evaluated this indicator before and after the implementation of the Picture Archiving and Communication system (PACS). Figure 7 shows that the interval was reduced from 30 to 22 days with a reduction of 25% of the average waiting time, proving the effectiveness of the process.

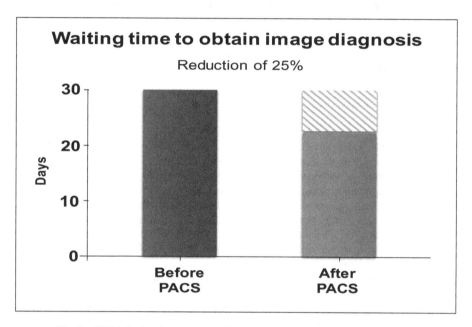

Fig. 7. INCA Patient's average waiting time to obtain CT image diagnosis.

5   Conclusions

Nowadays, healthcare organizations generate huge quantities of data, but regrettably, this asset is not yet entirely used for improving the management and delivery of healthcare services. The benefits gained from the implementation of this knowledge management framework are real-time knowledge access; knowledge share; costs reduction; cancer diagnosis agility and treatment quality.

The adoption of an integrated simulation environment can improve the efficiency of healthcare delivery allied with capturing and sharing patient data among different health professionals. The growth of simulation experiments over the last years has been encouraged by a great number of scientists and researchers conducting exploratory research with simulation models applied to diverse cases of healthcare services management.

This paper indicated how decision making in a cancer treatment center can be improved using a knowledge-based simulation framework. The use of this environment provided the necessary analytic support and insights into such operational decisions. The chosen approach was supported by the suite of applications which has retrieved information about the clinical history of patients, identified process bottlenecks and used discrete simulation technique to investigate alternative scenarios to reduce the patients' waiting time to realize CT's exams. The alternative which has indicated the PACS implementation has reduced patients' waiting time for cancer treatment. Its implementation showed, therefore, that good simulation experiments can make positive changes.

A clear understanding of the decision-making process by managers and clinical staff was essential. The reaction against new analytics tools on part of some physicians has been overcome by the development of user-friendly simulation interfaces. Another incentive has been the warm adhesion of young medical staff. A trend related to the use of graphics interfaces as a deployment critical success factor was observed.

References

1. Pronovost, P.J., Goeschel, C.A.: Viewing healthcare delivery as science: challenges, benefits, and policy implications. Health Serv. Res. **45**, 1508–1522 (2010)
2. Ali, N., Tretiakov, A., Whiddett, D., Hunter, I.: Knowledge management systems success in healthcare: leadership matters. Int. J. Med. Inform. **97**, 331–340 (2017)
3. Glasziou, P., Ogrinc, G., Goodman, S.: Can evidence-based medicine and clinical quality improvement learn from each other? BMJ Qual. Saf. **20**(Suppl. 1), i13–i17 (2011)
4. Ali, N., Whiddett, D., Tretiakov, A., Hunter, I.: The use of information technologies for knowledge sharing by secondary healthcare organizations in New Zealand. Int. J. Med. Inform. **81**(7), 500–506 (2012)
5. INCA - Instituto Nacional de Câncer José de Alencar Gomes da Silva. http://www.inca.gov.br/estimativa/2016
6. Goldblatt, E.M., Lee, W.H.: From bench to bedside: the growing use of translational research in cancer medicine. Am. J. Transl. Res. **2**(1), 1–18 (2010)
7. Eldabi, T.: Implementation issues of modeling healthcare problems: misconceptions and lessons. In: Proceedings of the 2009 Winter Simulation Conference, pp. 1831–1839 (2009)
8. Rosenhead, J., Mingers, J.: Rational Analysis for a Problematic World Revisited. Wiley, Chichester (2001)
9. VanBerkel, P.T., Blake, J.T.: A comprehensive simulation for wait time reduction and capacity planning applied in general surgery. Health Care Manag. Sci. **10**(4), 373–385 (2007)
10. Perugini, D., Perugini, M.: Characterized and personalized predictive-prescriptive analytics using agent-based simulation. Int. J. Data Anal. Tech. Strat. (IJDATS) **6**(3), 209–227 (2014)
11. Kershaw, R.: Using TOC to cure healthcare problems. Manag. Account. Q. (Spring), 22–28 (2000)

Data Science and Decision Systems
in Medicine

Part II: Data Science and Decision Systems in Medicine

The second part of this volume comprises the extended and revised versions of a set of selected work from the track "Data Science and Decision Systems in Medicine".

In the first paper of the second part, "Lifted Maximum Expected Utility", Gehrke *et al.* present an extension to a parameterised probabilistic model known as the lifted junction tree (LJT) algorithm for querying relational models under uncertainty, such as found in electronic health records (EHRs). This algorithm answers multiple queries efficiently for relational models under uncertainty by building then reusing a first-order cluster representation. By calculating a lifted solution to the maximum expected utility (MEU) problem, the underlying LJT model can be extended into a parameterised probabilistic decision model by adding action and utility nodes, thus creating the meuLJT algorithm, capable of solving the MEU problem using parameterised probabilistic decision models efficiently, while also being able to answer multiple marginal queries.

Next, "The Role of Usability Engineering in the Development of an Intelligent Decision Support System", Martin *et al.* describe the role of human factors and user-centred design in the creation of medical systems that adhere to international standards. After introducing the reader to the usability engineering process, the paper describes how human factors are being considered throughout the development of a personalised clinical decision support system for the management of type 1 diabetes called PEPPER, which applies artificial intelligence methods to recommend medication dosage. The design of the system involves users at every stage of the development and uses multiple techniques to maximise usability, minimise errors, and increase acceptance of recommendations by users. The preliminary analysis of data shows promising results.

The third paper in this part, "Automated Pain Detection in Facial Videos of Children Using Human-Assisted Transfer Learning", by Xu *et al.* tackles the task of automatically and accurately detecting pain in children from their facial expressions. Computer vision algorithms have been trained in the past to detect facial action units (AUs) and classification systems have been built to differentiate between pain and no pain, however these systems show variable performance depending on environmental factors. The paper presents an improvement in pain detection by applying transfer learning between automated AU codings and a subspace of manual AU codings to enable more robust pain recognition performance than when only automatically coded AUs are available for the test data. Using the transfer learning method, the authors improved the area under the ROC curve on an independent test set from 0:67 to 0:72.

In the fourth chapter of this part, "Towards Automated Pain Detection in Children Using Facial and Electrodermal Activity", Xu *et al.* apply a different method to automatically and accurately detect pain in children. In addition to the facial activity recognition, electro-dermal activity (EDA) is jointly exploited to build a more robust model through model fusion. The paper discusses the preliminary steps towards fusing models trained on video for facial activity recognition and on EDA respectively. The authors then compare fusion models using original video features and those using transferred video features, which are less sensitive to environmental changes. The paper concludes by demonstrating the advantage of the fusion between the transferred video features and the EDA features through improved performance relative to using EDA and video features alone.

Next, chapter "Interpretation of Best Medical Coding Practices by Case-Based Reasoning, A User Assistance Prototype for Data Collection for Cancer Registries", Schnell *et al.* highlight the importance of cancer registries as an important tool to fight this disease. At the heart of these registries lies the data collection and coding process. This process is ruled by complex international standards and a number of best practices, which can easily overwhelm (coding) operators. This paper presents a system assisting operators in the interpretation of best medical coding practices and its evaluation. By leveraging the arguments used by the coding experts to determine the best coding option, the proposed system answers coding questions from operators and provides a partial explanation for the proposed solution.

Chapter "Identification of Serious Illness Conversations in Unstructured Clinical Notes Using Deep Neural Networks", Chien *et al.* focus on applying deep learning to care planning, in an attempt to clarify and document goals of care and preferences for future care in the context of end-of-life care. To remain consistent with the preferences of dying patients and their families, physicians document their communication about these preferences as unstructured free text in clinical notes; as a result, routine assessment of this quality indicator is time consuming and costly. In this study, the authors trained and validated a deep neural network to detect documentation of advanced care planning conversations in clinical notes from electronic health records. The system performance was assessed against rigorous manual chart review and rule-based regular expressions. For detecting documentation of patient care preferences at the note level, the algorithm showed high F1 score, sensitivity and specificity, and consistently outperformed regular expressions.

Chapter "Generating Reward Functions using IRL Towards Individualized Cancer Screening", Petousis *et al.* demonstrate how inverse reinforcement learning (IRL) can serve precision medicine by providing individualized cancer screening recommendations and consequently decreasing overdiagnosis. Personalized cancer screening addresses the heterogeneity of cancer screening participants. Partially observable Markov decision processes (POMDPs), when defined with an appropriate reward function, can be used to suggest optimal, individualized screening policies. However, determining an appropriate reward function can be challenging. This paper proposes the use of inverse reinforcement learning to form rewards functions for lung and breast cancer screening POMDPs. Using experts (physicians) retrospective screening decisions for lung and breast cancer screening, the paper presents two POMDP models with corresponding reward functions, namely the maximum entropy (MaxEnt) IRL algorithm with an adaptive step size and the multiplicative model to learn state-action pair rewards. These reward functions, when combined with POMDP models in lung and breast cancer screening, demonstrate a performance comparable to that of experts.

Finally, chapter "Deep Learning Architectures for Vector Representations of Patients and Exploring Predictors of 30-Day Hospital Readmissions in Patients with

Multiple Chronic Conditions", Rafiq *et al.* explore the usefulness of deep learning architectures to identify patient segments and factors contributing to 30-day hospital readmissions in a group of patients affected by complex chronic concurrent conditions (diabetes, cardiovascular and kidney diseases). A convolutional neural network (CNN) and recurrent neural network (RNN) was implemented on sequential electronic health records data at the Danderyd Hospital in Stockholm, Sweden. Three distinct subtypes of patient groups were identified: chronic obstructive pulmonary disease, kidney transplant, and paroxysmal ventricular tachycardia. The CNN learned vector representations of patients, but the RNN was better able to identify and quantify key contributors to readmission such as myocardial infarction and echocardiography. This study suggests that a vector representation of patients with deep learning should precede predictive modelling of complex patients. The approach also has potential implications for supporting care delivery, care design and clinical decision-making.

Lifted Maximum Expected Utility

Marcel Gehrke[1]([⊠])(iD), Tanya Braun[1](iD), Ralf Möller[1], Alexander Waschkau[2],
Christoph Strumann[2], and Jost Steinhäuser[2]

[1] Institute of Information Systems, University of Lübeck, Lübeck, Germany
{gehrke,braun,moeller}@ifis.uni-luebeck.de
[2] Institute of Family Medicine, University Medical Center Schleswig-Holstein,
Campus Lübeck, Lübeck, Germany
{alexander.waschkau,christoph.strumann,jost.steinhaeuser}@uksh.de

Abstract. The lifted junction tree algorithm (LJT) answers multiple
queries efficiently for relational models under uncertainties by building and then reusing a first-order cluster representation. We extend
the underling model representation of LJT, which is called parameterised probabilistic model, to calculate a lifted solution to the maximum
expected utility (MEU) problem. Specifically, this paper contributes (i)
action and utility nodes for parameterised probabilistic models, resulting in parameterised probabilistic decision models and (ii) meuLJT, an
algorithm to solve the MEU problem using parameterised probabilistic decision models efficiently, while also being able to answer multiple
marginal queries.

1 Introduction

Areas such as health care and logistics involve probabilistic data with relational
aspects and need efficient exact inference algorithms, which allow for decision
support. These areas involve many objects in relation to each other with uncertainties about object existence, attribute value assignments, or relations between
objects. More specifically, health care systems involve electronic health records
(EHRs) (the relational part) for many patients (the objects) and uncertainties
[18] due to, e.g., missing information caused by data integration from different
hospitals or faulty sensors. Automatically analysing EHRs can improve the care
of patients and save time. In this paper, we study the problem of exact decision
making under uncertainty in lifted probabilistic models.

Braun and Möller [2] investigate parameterised probabilistic models (PMs) to
represent probabilistic relational behaviour, and furthermore introduce the lifted
junction tree algorithm (LJT), an exact inference algorithm to answer multiple
queries efficiently. Specifically, this paper contributes (i) action and utility nodes
for parameterised probabilistic models, resulting in parameterised probabilistic
decision models (PDecMs) and (ii) meuLJT, an algorithm to solve the maximum

This research originated from the Big Data project being part of Joint Lab 1, funded
by Cisco Systems Germany, at the centre COPICOH, University of Lübeck.

F. Koch et al. (Eds.): AIH 2018, LNAI 11326, pp. 131–141, 2019.
https://doi.org/10.1007/978-3-030-12738-1_10

expected utility (MEU) problem using PDecMs efficiently, while also being able to answer multiple marginal queries.

Action nodes are well-motivated candidates to model, e.g., treatments, while utility nodes can represent, e.g., the well being of patients, risk scores, or treatment costs. With utilities modelling is not restricted to a single particular area, but one can also model a combination of areas, such as well being of patients and risk scores.

Health care needs exact results as approximations might not be good enough [19]. Further, the lifting approach exploits symmetries in the model to reduce the number of instances or patients to perform inference on. Additionally, LJT clusters a model into submodels to efficiently answer queries, like the condition of each patient. Therefore, LJT is suitable to handle health care related data.

In the following, we recapitulate PMs as a representation for relational probabilistic models and introduce PDecMs, by adding actions and utilities to the representation. Afterwards, we formalise the MEU problem and discuss different modelling possibilities, also from an ethical point of view. Lastly, we introduce meuLJT to reuse computations and answer multiple queries efficiently.

2   Related Work

We take a look at inference under uncertainty in relational models as well as relational decision support.

First-order probabilistic inference leverages the relational aspect of a static model. For models with known domain size, it exploits symmetries in a model by combining instances to reason with representatives, known as lifting [11]. Poole [11] introduces parametric factor graphs as relational models and proposes lifted variable elimination (LVE) as an exact inference algorithm on relational models. Further, de Salvo Braz [12], Milch et al. [7], and Taghipour et al. [17] extend LVE to its current form. Lauritzen and Spiegelhalter [6] introduce the junction tree algorithm. To benefit from the ideas of the junction tree algorithm and LVE, Braun and Möller [2] present LJT, which efficiently performs exact first-order probabilistic inference on relational models given a set of queries.

Nath and Domingos [8] introduce Markov logic decision networks (MLDNs), which are relational static models with action and utility nodes. Nath and Domingos calculate approximate solutions to the static MEU problem in a completely grounded way [10] based on MLDNs. Another approach of Nath and Domingos include unnecessary groundings [9]. Further, Apsel and Brafman [1] propose an exact lifted solution to the MEU problem based on [8]. These approaches are designed to handle single queries. However, we propose to answer multiple queries efficiently.

Additional research focuses on sequential decision making by investigating first-order (partially observable) Markov decision processes (FO (PO)MDPs) [5,14,15], which use lifting techniques from de Salvo Braz, Amir, and Roth [13]. In contrast to FO POMDPs, which perform offline policy iteration, we propose to support probabilistic online planning.

3   Parameterised Probabilistic Models

Based on [4], we recapitulate PMs for relational probabilistic models. PMs combine first-order logic with probabilistic models, representing first-order constructs using logical variables (logvars) as parameters. Let us assume, we would like to remotely infer the condition of patients with regards to water retaining. To determine the condition of patients, we use the change of their weights. An increase in weight could either be caused by overeating or retaining water. Additionally, we use the change of weights of people living with the patient to reduce the uncertainty to infer conditions. In case both persons gain weight, overeating is more likely, while otherwise retaining water is more likely. If a water retention is undetected, it can be an acute life-threatening condition.

People behave in the same way w.r.t. gaining weight if we are interested whether a person retains water. For a water retention, persons gain weight over a few days in a way which would be hard to achieve by overeating each day. Thus, if we are interested whether they retain water, having information about the weight gain of persons is independent of the actual person. Hence, we can have a random variable (randvar) for each person about their current condition. As persons behave the same w.r.t. gaining weight and PMs allow for using logvars as parameters, we can construct a parameterised randvar (PRV) with the persons as logvar for our randvar.

Definition 1. *Let **L** be a set of logvar names, Φ a set of factor names, and **R** a set of randvar names. A PRV $A = P(X^1, ..., X^n)$ represents a set of randvars behaving identically by combining a randvar $P \in \mathbf{R}$ with $X^1, ..., X^n \in \mathbf{L}$. If $n = 0$, the PRV is parameterless. The domain of a logvar L is denoted by $\mathcal{D}(L)$. The term range(A) provides possible values of a PRV A. Constraint $(\mathbf{X}, C_{\mathbf{X}})$ allows to restrict logvars to certain domain values and is a tuple with a sequence of logvars $\mathbf{X} = (X^1, ..., X^n)$ and a set $C_{\mathbf{X}} \subseteq \times_{i=1}^n \mathcal{D}(X^i)$. $\top$ denotes that no restrictions apply and may be omitted. The term $lv(Y)$ refers to the logvars and $rv(Y)$ to the randvars in some element Y. The term $gr(Y|C)$ denotes the set of instances of Y with all logvars in Y grounded w.r.t. constraint C.*

To model our scenario, we use the randvar names C, LT, S, and W for Condition, LivingTogether, ScaleWorks, and Weight, respectively, and the logvar names X and X'. From the names, we build PRVs $C(X)$, $LT(X, X')$, $S(X)$, and $W(X)$. The domain of X and X' is $\{alice, bob, eve\}$. The range of $C(X)$ is $\{normal, deviation, retains\ water\}$. $LT(X, X')$ and $S(X)$ have range $\{true, false\}$ and $W(X)$ has range $\{steady, falling, rising\}$. A constraint $C = (X, \{alice, bob\})$ for X allows for restricting X to a subset of its domain, in this case to $alice$ and bob. Using the constraint, the expression $gr(W(X)|C)$ evaluates to $\{W(alice), W(bob)\}$. The expression $gr(W(X)|\top)$ also contains $W(eve)$. Now, we define parametric factors (parfactors), to set PRVs into relation to each other.

Definition 2. *We denote a parfactor g with $\forall \mathbf{X} : \phi(\mathcal{A}) \mid C$. $\mathbf{X} \subseteq \mathbf{L}$ being a set of logvars over which the factor generalises and $\mathcal{A} = (A^1, ..., A^n)$ a sequence of*

PRVs. We omit ($\forall \mathbf{X}$:) if $\mathbf{X} = lv(\mathcal{A})$. A function $\phi : \times_{i=1}^{n} range(A^i) \mapsto \mathbb{R}^+$ with name $\phi \in \Phi$ is defined identically for all grounded instances of $\mathcal{A}$. A list of all input-output values is the complete specification for ϕ. C is a constraint on $\mathbf{X}$. A PM $G := \{g^i\}_{i=0}^{n}$ is a set of parfactors and semantically represents the full joint probability distribution $P_G = \frac{1}{Z} \prod_{f \in gr(G)} f$ where Z is a normalisation constant.

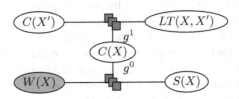

Fig. 1. Parfactor graph for G^{ex}, the weight is observable

Now, we build the model G^{ex} of our example with the parfactors:

$$g^0 = \phi^0(C(X), S(X), W(X))|\top \text{ and } g^1 = \phi^1(C(X), C(X'), LT(X, X'))|\kappa^1$$

We omit the concrete mappings of ϕ^0 and ϕ^1. Parfactor g^0 has the constraint $\top$, meaning it holds for *alice*, *bob*, and *eve*. The constraint κ^1 of g^1 ensures that $X \neq X'$ holds. Figure 1 depicts G^{ex} as a parfactor graph and shows PRVs, which are connected via undirected edges to parfactors, with $W(X)$ being observable. We can observe the weight of patients. The remaining PRVs are latent.

The semantics of a model is given by grounding and building a full joint distribution. In general, queries ask for a probability distribution of a randvar using a model's full joint distribution and fixed events as evidence.

Definition 3. *Given a PM G, a query term Q (ground PRV), and events $\mathbf{E} = \{E^i = e^i\}_i$ (ground PRVs with fixed range values), the expression $P(Q|\mathbf{E})$ denotes a query w.r.t. P_G.*

In our example, a query is $P(C(bob)|W(bob) = steady)$, asking for the probability distribution of *bob*'s condition given information about his weight.

4 Lifted Maximum Expected Utility

In this section, we introduce actions and utilities to PMs and show how to solve the MEU problem, by formalising the problem. Further, we discuss different modelling possibilities with PDecMs.

4.1   Parameterised Probabilistic Decision Models

Let us extend PMs with action and utility nodes, resulting in PDecMs.

Definition 4. *We represent actions and utilities by PRVs. Let Φ^u be a set of utility factor names. The range of action PRVs is disjoint actions and the range of utility PRVs is $\mathbb{R}$. A parfactor that maps to a utility PRV U is a utility par-factor. We denote a utility parfactor u with $\forall \mathbf{X} : \mu(\mathcal{A}) \,|C$, where C a constraint on $\mathbf{X}$. Function $\mu : \times_{i=1}^{n-1} range(A^i) \mapsto \mathbb{R}$, $A^i \in \mathcal{A}$, with name $\mu \in \Phi^u$ is defined identically for all grounded instances of $\mathcal{A}$ and its output is the value of U. A PDecM G is a PM with an additional set G^u of utility parfactors. Let $rv(G^u)$ refer to all probability randvars in G^u. Then, G^u semantically represents the combination of all utilities $U_G = \sum_{f \in gr(G^u)} f$.*

The μ functions output a utility, i.e., a scalar, which makes comparing utility values easy. Further, a scalar allows for testing whether utilities are within an ϵ margin of each other, making them hardly discriminable. With utilities incorpo-rated, we look at actions. To model actions, we introduce an action PRV with the actions in its range. Hence, we have one PRV, which models disjoint actions. To execute an action, we set the value of the action PRV to the action, which we want to perform, similar to providing evidence for marginal queries. Thus, the range of an action PRV $A(X)$ consists of different actions, lets say $A^1, ..., A^n$, and by setting $A(X)$ to the action, lets say A^1 ($A(X) = A^1$), we can select the action we would like to perform

Let us now extend the example with action and utility nodes. In Fig. 2, one can see one action node (square), one utility node (diamond), and one utility parfactor (crosses). In our example, the action PRV $A(X)$ has two actions in its range, namely A^1 is visit patient and action A^2 is do nothing. Obviously, other actions could also be included in the model, e.g., diet related actions or obtaining a more accurate scale.

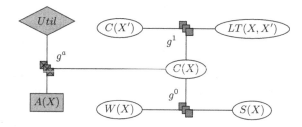

Fig. 2. Retaining water example with action and utility nodes in grey

In our example, the condition of patients and A^1 influence the utility. For example, patients with a chronic heart failure might tend to retain water. In case water retention is detected early on, treatment is easier. However, if this water retention remains undetected, water can also retain in the lung, which can lead to a pulmonary edema, making a treatment more costly. More importantly, pulmonary edema is an acute life-threatening condition. In addition to the condition of patients, A^1 also influences the utility as a doctor, with limited time, visiting a patient is expensive. Thus, one always needs to consider that alerting the doctor too early generates unnecessary costs and alerting the doctor too late can have serious consequences for the patient.

4.2 Maximum Expected Utility

To select the best action, we define queries and expected utility on a PDecM.

Definition 5. *Given a PDecM G, a query term Q, and events $\mathbf{E}$, the expression $P(Q|\mathbf{E}, \mathbf{s})$ denotes a probability query w.r.t. P_G. Given an assignment $\mathbf{a}$ for actions, the expression $U(Q, \mathbf{E}, \mathbf{a})$ refers to a utility w.r.t. U_G. The expected utility of G is defined by*

$$eu(G|\mathbf{a}) = \sum_{v \in range(rv(G^u))} P(v|\mathbf{a}) \cdot U(v, \mathbf{a}) \tag{1}$$

The inner part of the summation in Eq. (1) calculates a belief state $P(v|\mathbf{a})$ and combines it with corresponding utilities $U(v, \mathbf{a})$. By summing over all randvars from G^u, one obtains a scalar representing the expected utility. LVE allows for exactly computing an expected utility. Based on expected utility, we define the MEU as follows.

Definition 6. *Given a PDecM G, the MEU problem is given by*

$$meu[G] = \left(\arg\max_{\mathbf{a}} eu(G|\mathbf{a}), \max_{\mathbf{a}} eu(G|\mathbf{a}) \right) \tag{2}$$

Equation (2) suggest a naive algorithm defining how to calculate the MEU, namely by iterating over all possible action configurations, computing an expected utility for each configuration using LVE, an iteration that one cannot avoid if asking for an exact solution. The action assignment that maximises the expected utility is selected. As the utility value is a scalar, the expected utility w.r.t. configurations can be easily compared. Therefore, we also can easily determine configurations whose expected utility lie within an ϵ margin. In case different actions lie within an ϵ margin, the actions are hardly discriminable w.r.t. utilities.

The action PRV in G^{ex} has two possible actions. By setting $A(X) = A1$, we turn on A^1. By setting $A(X) = A2$, we turn on A^2. Thus, in our example to calculate the MEU, we need to iterate over two action assignments. For each expected utility, we obtain a scalar, allowing us to easily compare them and return the action with the MEU and the actual expected utility value. If all patients behave the same, we only need to iterate over two actions. In case we obtain different evidence for lets say two groups of patients, X^1 and X^2, we need to iterate over the actions for both groups. Hence, we would need to iterate over $\{A(X^1) = A1, A(X^2) = A1\}$, $\{A(X^1) = A2, A(X^2) = A1\}$, $\{A(X^1) = A1, A(X^2) = A2\}$, and $\{A(X^1) = A2, A(X^2) = A2\}$. In general, we need to iterate over r^n actions, where r is the number of actions in the range of an action PRV and n the number of different groups. Assuming, we have ten patients in two groups and two possible actions. Solving the MEU in a lifted way, we need to iterate over $2^2 = 4$ actions. Without the lifting idea, we would need to iterate over $2^{10} = 1024$ actions. Therefore, solving the MEU problem in a lifted way makes the problem manageable.

4.3   How to Model Utilities in a Medical Context

For decision support in a medical context, the model has to take into account the prevalence, i.e., the probability, of the diseases or health related problems to be identified. The prevalence does not only depend on the value of data but also on the source of data. For example, to identify a coronary heart diseases the prevalence is higher if the data comes from a chest pain unit compared to examinations from general practice [16]. In this context, the knowledge of the sensitivity and specificity of the analytical model and the prevalence is very important. Ideally, the model should inform the physician about its sensitivity and specificity to clarify the probability of a false positive result for each patient regarding the pre/post test probability. These information can help to plan further treatment and diagnostic decisions. The aim of the model should be to avoid unnecessary examinations and thus costs. Further, decision making should not unsettle the patient on the one hand, but on the other hand detect serious conditions timely.

As PDecMs can model different influences, we can take prevalence into account. Thus, we need to model different PRVs for different sources, which then depending on the value of the test results, having different impact on the condition of a patient. Further, there are two different kinds of queries for PDecMs, namely utility and probability queries. Thus, we can also state marginal queries. Having marginal queries, we can also query the current belief of the condition of a patient as well as the condition of a patient after an action, i.e., treatment or test, is performed.

5   Solving the MEU Problem and Answer Multiple Marginal Queries Efficiently

In this section, we recapitulate LJT [3] to answer queries for PMs and introduce meuLJT to solve the MEU problem and answer multiple marginal queries using PDecMs efficiently.

5.1   Lifted Junction Tree Algorithm

LJT provides efficient means to answer queries $P(Q^i|\mathbf{E})$, with $Q^i \in \mathbf{Q}$ a set of query terms, given a PM G and evidence $\mathbf{E}$, by performing the following steps:

(i) Construct an first-order junction tree (FO jtree) J for G.
(ii) Enter $\mathbf{E}$ in J.
(iii) Pass messages.
(iv) Compute answer for each query $Q^i \in \mathbf{Q}$.

We first define an FO jtree and then go through each step. To define an FO jtree, we define parameterised clusters (parclusters), nodes of an FO jtree.

Definition 7. *A parcluster* $\mathbf{C}$ *is defined by* $\forall \mathbf{L} : \mathbf{A} | C$. $\mathbf{L}$ *is a set of logvars,* $\mathbf{A}$ *is a set of PRVs with* $lv(\mathbf{A}) \subseteq \mathbf{L}$, *and* C *a constraint on* $\mathbf{L}$. *We omit* $(\forall \mathbf{L} :)$ *if* $\mathbf{L} = lv(\mathbf{A})$. *A parcluster* $\mathbf{C}^i$ *can have parfactors* $\phi(\mathcal{A}^\phi) | C^\phi$ *assigned given that (i)* $\mathcal{A}^\phi \subseteq \mathbf{A}$, *(ii)* $lv(\mathcal{A}^\phi) \subseteq \mathbf{L}$, *and (iii)* $C^\phi \subseteq C$ *holds. We call the set of assigned parfactors a local model* G^i.
An FO jtree for a PM G *is* $J = (\mathbf{V}, \mathbf{P})$ *where* J *is a cycle-free graph, the nodes* $\mathbf{V}$ *denote a set of parclusters, and* $\mathbf{P}$ *is a set of edges between parclusters.* J *must satisfy the following properties: (i) A parcluster* $\mathbf{C}^i$ *is a set of PRVs from* G. *(ii) For each parfactor* $\phi(\mathcal{A})|C$ *in* G, $\mathcal{A}$ *must appear in some parcluster* $\mathbf{C}^i$. *(iii) If a PRV from* G *appears in two parclusters* $\mathbf{C}^i$ *and* $\mathbf{C}^j$, *it must also appear in every parcluster* $\mathbf{C}^k$ *on the path connecting nodes* i *and* j *in* J *(running intersection). The separator* $\mathbf{S}^{ij}$ *of edge* $i - j$ *is given by* $\mathbf{C}^i \cap \mathbf{C}^j$ *containing shared PRVs.*

LJT constructs an FO jtree using a first-order decomposition tree, enters evidence in the FO jtree, and to distribute local information of the nodes through the FO jtree, passes messages through an *inbound* and an *outbound* pass. To compute a message, LJT eliminates all non-separator PRVs from the parcluster's local model and received messages. After message passing, LJT answers queries. For each query, LJT finds a parcluster containing the query term and sums out all non-query terms in its local model and received messages.

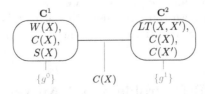

Fig. 3. FO jtree for G^{ex} (local models under the parclusters)

Figure 3 shows an FO jtree of G^{ex} with the local models of the parclusters and the separators as labels of edges. During the *inbound* phase of message passing, LJT sends messages from $\mathbf{C}^1$ to $\mathbf{C}^2$ and for the *outbound* phase a message from $\mathbf{C}^2$ to $\mathbf{C}^1$. If we would like to know whether $S(bob)$ holds, we query $P(S(bob))$ for which LJT can use parcluster $\mathbf{C}^1$. LJT sums out $C(X)$, $W(X)$, and $S(X)$ where $X \neq bob$ from $\mathbf{C}^1$'s local model G^1, $\{g^0\}$, combined with the received messages.

5.2   meuLJT

Now, we introduce meuLJT to solve the MEU problem. For now, we restrict a PDecM G to have at most one utility PRV and one utility parfactor. The basic step of meuLJT are similar to LJT, namely:

(i) Construct an FO jtree J for G.
(ii) Enter evidence and actions in J.
(iii) Pass messages.
(iv) Compute answer queries.

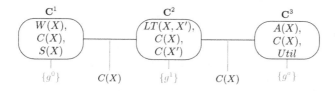

Fig. 4. FO jtree for the PDecM G^{ex} (local models under the parclusters)

Let us now develop an idea about how meuLJT solves the lifted MEU problem. Figure 4 shows an FO jtree for the PDecM G^{ex}. Compared to the FO jtree from Fig. 3, we see an additional parcluster $\mathbf{C}^3$ with the utility parfactor. To construct the FO jtree, meuLJT treats the utility parfactor as any other parfactor. Including utility parfactors in the parcluster definition is straight forward. Using the FO jtree, meuLJT distributes local information by message passing. To calculate the probability messages, meuLJT also performs a message pass. During the message pass, meuLJT excludes utility parfactors as they do not influence the probability distributions and we only have one utility parfactor and one utility PRV. Hence, during the *inbound* pass, $\mathbf{C}^2$ receives a message over $C(X)$ from $\mathbf{C}^1$ and an empty message from $\mathbf{C}^3$. For the *outbound* pass $\mathbf{C}^2$ sends messages over $C(X)$ to $\mathbf{C}^1$ and $\mathbf{C}^3$. To calculate utilities, utility parfactors need to know the probability distributions, which is distributed by message passing also to parclusters with utility parfactors. Now, meuLJT can use $\mathbf{C}^1$ and $\mathbf{C}^2$ to answer marginal queries and $\mathbf{C}^3$ to answer *expected utility* queries. Given new evidence or a new action assignment meuLJT has to recompute messages. Hence, for each action assignment meuLJT can answer the *expected utility* query and efficiently answer multiple marginal queries, e.g., of the condition of patients.

In our example, we have two action sequences $\{A(X) = A1\}$ and $\{A(X) = A2\}$, if all patients behave the same. To calculate the MEU, meuLJT has to iterate over all action sequences and calculate the corresponding expected utility. For the first action sequence, meuLJT enters $\{A(X) = A1\}$ as evidence in the FO jtree from Fig. 4. After the message pass, meuLJT uses $\mathbf{C}^3$ to answer the *expected utility* query for action $\{A(X) = A1\}$. $\mathbf{C}^3$ received the current belief state during message passing and has the current action due to the evidence. Thus, all required information to calculate the expected utility are present.

For the second action sequence, meuLJT enters $\{A(X) = A2\}$ as evidence in the FO jtree from Fig. 4. Normally meuLJT would need to perform a new message pass, but the evidence does not change any calculations of the probability messages in this case. Thus, meuLJT can reuse the already performed message pass. Hence, meuLJT can directly use $\mathbf{C}^3$ to answer the *expected utility* query for action $\{A(X) = A2\}$. $\mathbf{C}^3$ received the current belief state during message passing and has the current action due to the evidence. Thus, all required information to calculate the expected utility are present. Having the expected utility for both actions, meuLJT selects the action with the MEU. In case, we have more actions or have more groups of patients, meuLJT has more action sequences to iterate over. In general, as long as we only have one action PRV and one utility PRV, and both occur only in a utility parfactor, meuLJT can reuse the message pass and thereby, prevent redundant calculations.

All in all, meuLJT directly reasons over all patients instead of reason over each patient individually. Additionally, meuLJT can provide alerts based on observations of each patient. Apsel and Brafman [1] extend C-FOVE to solve MEU queries, which significantly outperforms the propositional case. Braun and Möller [2] show that LJT outperforms GC-FOVE, an extension to C-FOVE, for multiple queries. Therefore, meuLJT is well-suited to support lifted decision making and answering multiple marginal queries.

6   Conclusion

We present meuLJT to support lifted decision making by calculating a solution to the MEU problem efficiently. Areas like health care benefit from the lifting idea for many patients and the support of different kinds of queries. By extending the underlying model with action and utility nodes, complete health care processes including treatments can be modelled. Additionally, by maximising the expected utility, meuLJT can calculate the best action.

The next step is to extend meuLJT and the underlying problem to the temporal case. Further, we investigate whether, for our application, evidence can reduce the MEU problem roughly from a POMDP to an MDP.

References

1. Apsel, U., Brafman, R.I.: Extended lifted inference with joint formulas. In: Proceedings of the 27th Conference on Uncertainty in Artificial Intelligence, pp. 11–18. AUAI Press (2011)
2. Braun, T., Möller, R.: Lifted junction tree algorithm. In: Friedrich, G., Helmert, M., Wotawa, F. (eds.) KI 2016. LNCS (LNAI), vol. 9904, pp. 30–42. Springer, Cham (2016). https://doi.org/10.1007/978-3-319-46073-4_3
3. Braun, T., Möller, R.: Parameterised queries and lifted query answering. In: IJCAI, pp. 4980–4986 (2018)

4. Gehrke, M., Braun, T., Möller, R.: Lifted dynamic junction tree algorithm. In: Chapman, P., Endres, D., Pernelle, N. (eds.) ICCS 2018. LNCS (LNAI), vol. 10872, pp. 55–69. Springer, Cham (2018). https://doi.org/10.1007/978-3-319-91379-7_5

5. Joshi, S., Kersting, K., Khardon, R.: Generalized first order decision diagrams for first order Markov decision processes. In: IJCAI, pp. 1916–1921 (2009)

6. Lauritzen, S.L., Spiegelhalter, D.J.: Local computations with probabilities on graphical structures and their application to expert systems. J. Roy. Stat. Soc. Ser. B (Methodol.) 50(2), 157–224 (1988)

7. Milch, B., Zettlemoyer, L.S., Kersting, K., Haimes, M., Kaelbling, L.P.: Lifted probabilistic inference with counting formulas. In: Proceedings of AAAI, vol. 8, pp. 1062–1068 (2008)

8. Nath, A., Domingos, P.: A language for relational decision theory. In: Proceedings of the International Workshop on Statistical Relational Learning (2009)

9. Nath, A., Domingos, P.: Efficient lifting for online probabilistic inference. In: Proceedings of the Twenty-Fourth AAAI Conference on Artificial Intelligence, pp. 1193–1198. AAAI Press (2010)

10. Nath, A., Domingos, P.M.: Efficient belief propagation for utility maximization and repeated inference. In: Proceedings of the Twenty-Fourth AAAI Conference on Artificial Intelligence, pp. 1187–1192. AAAI Press (2010)

11. Poole, D.: First-order probabilistic inference. In: Proceedings of IJCAI, vol. 3, pp. 985–991 (2003)

12. de Salvo Braz, R.: Lifted first-order probabilistic inference. Ph.D. thesis, Ph. D. dissertation, University of Illinois at Urbana Champaign (2007)

13. de Salvo Braz, R., Amir, E., Roth, D.: MPE and partial inversion in lifted probabilistic variable elimination. In: AAAI, vol. 6, pp. 1123–1130 (2006)

14. Sanner, S., Boutilier, C.: Approximate solution techniques for factored first-order MDPs. In: 17th International Conference on Automated Planning and Scheduling, ICAPS 2007, pp. 288–295. AAAI Press (2007)

15. Sanner, S., Kersting, K.: Symbolic dynamic programming for first-order POMDPs. In: Proceedings of the Twenty-Fourth AAAI Conference on Artificial Intelligence, pp. 1140–1146. AAAI Press (2010)

16. Steinhäuser, J., Kühlein, T.: Role of the general practitioner. In: Gombotz, H., Zacharowski, K., Spahn, D.R. (eds.) Patient Blood Management, pp. 61–65. Thieme, Stuttgart (2015)

17. Taghipour, N., Fierens, D., Davis, J., Blockeel, H.: Lifted variable elimination: decoupling the operators from the constraint language. J. Artif. Intell. Res. 47(1), 393–439 (2013)

18. Theodorsson, E.: Uncertainty in measurement and total error: tools for coping with diagnostic uncertainty. Clin. Lab. Med. 37(1), 15–34 (2017)

19. Wemmenhove, B., Mooij, J.M., Wiegerinck, W., Leisink, M., Kappen, H.J., Neijt, J.P.: Inference in the promedas medical expert system. In: Bellazzi, R., Abu-Hanna, A., Hunter, J. (eds.) AIME 2007. LNCS (LNAI), vol. 4594, pp. 456–460. Springer, Heidelberg (2007). https://doi.org/10.1007/978-3-540-73599-1_61

The Role of Usability Engineering in the Development of an Intelligent Decision Support System

Clare Martin$^{(\boxtimes)}$, Arantza Aldea , David Duce , Rachel Harrison ,
and Bedour Alshaigy$^{(\boxtimes)}$

School of Engineering, Computing and Mathematics, Oxford Brookes University,
Oxford, UK
{cemartin,aaldea,daduce,rachel.harrison,b.alshaigy}@brookes.ac.uk

Abstract. This paper presents an overview of the usability engineering process for the development of a personalised clinical decision support system for the management of type 1 diabetes. The tool uses artificial intelligence (AI) techniques to provide insulin bolus dose advice and carbohydrate recommendations that adapt to the individual. We describe the role of human factors and user-centred design in the creation of medical systems that must adhere to international standards. We focus specifically on the formative evaluation stage of this process. The preliminary analysis of data shows promising results.

Keywords: Artificial intelligence · Usability · Safety · Human factors

1 Introduction

In 2006, a patient accidentally received over four hours a dose of the chemotherapy drug fluorouracil that should have been administered over four days, and the consequences were fatal [21]. It has been argued [32] that this mistake was just one of many errors caused by usability issues with healthcare technology and the need for safety systems. In this case, the nurse had to determine the dose by a complex calculation, and the mental effort involved meant that the error was not detected.

Most people with type 1 diabetes (T1D) have to perform multivariate dose calculations several times a day, in a variety of contexts that might affect cognitive load. Many use mobile decision support tools to assist with the process [24,25], but these typically use simple formulae based on a limited set of parameters.

This paper describes how human factors are being considered in the design and evaluation of a more sophisticated system. PEPPER (Patient Empowerment through Predictive PERsonalised decision support) [16] is a tool that takes into account multiple parameters as input and uses artificial intelligence to offer dose advice. The design of the system involves users at every stage of the development

© Springer Nature Switzerland AG 2019
F. Koch et al. (Eds.): AIH 2018, LNAI 11326, pp. 142–161, 2019.
https://doi.org/10.1007/978-3-030-12738-1_11

to ensure that it meets their needs. It uses multiple techniques to maximise usability, minimise errors, and increase acceptance of recommendations by users.

1.1 Healthcare Technology Acceptance

Poor usability has long been identified as one of the barriers contributing to the lack of adoption of intelligent personal guidance systems for diabetes management [4,16]. Additional factors can affect the acceptance of healthcare technology more broadly [10] (see Fig. 1), even when benefits are backed by robust evidence. The attitude of patients impacts the perception of health care practitioners and also their willingness to accept innovations.

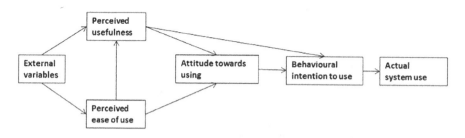

Fig. 1. Technology acceptance model [10].

There are related issues surrounding trust in applied artificial intelligence (AI). Public discussion of such topics is growing and not all perceptions are positive. There have been concerns regarding loss of control [12] and scepticism towards systems that replace human decision making [15] for example.

1.2 The PEPPER System

PEPPER is a research project funded by the European Union (EU) Horizon 2020 Framework [16]. It aims at creating a portable personalised decision support system, to provide bolus dose advice that is highly tailored towards the needs of individuals. The design has a dual architecture to accommodate treatment either by multiple daily injections (MDI) or continuous subcutaneous insulin infusion (CSII) through a patch-pump [9]. Most of the input is collected wirelessly, as shown in Fig. 2, by continuous glucose monitor (CGM), physical activity monitor and capillary glucose monitor. The latter is included to calibrate the CGM or in case CGM data is unavailable. Additional data such as food intake, alcohol consumption and hormonal cycles are input through the smartphone (for MDI) or a pump. Carbohydrate intake is the only mandatory manual input, because most people find such interactions tedious [3,25].

The information gathered by the handset is processed by a case-based reasoning [1] (CBR) module to determine a personalised insulin recommendation

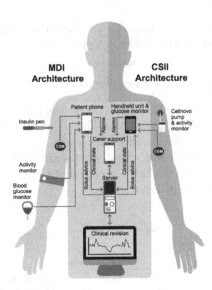

Fig. 2. The PEPPER system architecture.

that adapts over time, depending on outcomes. The key concept that underpins the PEPPER system is that CBR solves new problems in much the same way as a human does, based on past history, but with much greater accuracy, especially when informed by the diversity of data that can now be harvested via wearable technology. A second model-based reasoning (MBR) module is used to maximise safety [17]. The computer model generates predictive glucose alarms, automatic insulin suspension, carbohydrate recommendations, and fault diagnosis. The development of PEPPER uses an iterative methodology, integrated with clinical validation and formative usability evaluation. Methods for the latter are described in the next section.

1.3   User-Centred Design

The usability engineering process for medical software is more rigorous than that for other domains because of the need to consider safety and hazards. This requirement is encapsulated by standards such as International Electrotechnical Commission (IEC) 62366 [19], which is recognised by, and similar to, the guidance offered by the U.S. Food and Drug Administration (FDA) [34]. Both protocols emphasise the importance of conducting a user-centred design to determine tasks and frequently used functions, as well as identifying risks and use-related errors, prioritised by severity of harm. Some of the terminology varies however. For example, the terms "usability engineering" and "human factors engineering" are used interchangeably [20].

One of the shortcomings of the IEC standard is that it offers very little advice about how to evaluate technology in context, a crucial consideration for mobile devices. This is particularly applicable to the design and evaluation of medical devices, which might be used under adverse conditions. For example, a person with T1D may need to interact with technology when hypoglycaemic. Their interaction could therefore be affected by associated symptoms such as shakiness or blurred vision. One way to gain an understanding of people's experiences and the real scenarios of use is to employ "situated methods" [29]. Similar methods are therefore included in the protocol described here.

One feature that can play a crucial role in addressing the usability of intelligent systems is data visualisation. Research of its application to diabetes management is scarce [3]. Yet people with T1D need to be comfortable with interpreting mathematical data in order to manage their condition. An effective interface should allow people to feel in control for decisions such as whether to accept dose advice or not. We also conjecture that meaningful visual feedback helps developing positive health beliefs regarding the efficacy of AI to improve glycaemic outcomes.

1.4   Paper Scope

The contribution of this paper is to outline a usability evaluation protocol that adheres to the international standard for medical software [19]. Due to space limitations, we are unable to include the details of all stages of this protocol. We have therefore chosen to include the specifics of the method and results for the formative evaluation stage only, in Sects. 2.3 and 3.3 respectively. The remaining parts of Sects. 2 and 3 give an overview of the rest of the process in order to put the formative step into context. The results from these stages will be published separately.

2   Methods

Table 1 presents the usability engineering process used for PEPPER. The table is a refinement of the IEC 62366 standard [19]. As such, it comprises four stages each with its relevant steps. Definitions of the terms in uppercase can be found on the International Electrotechnical Commission Online Browsing Platform (OBP) [22] and the International Organization for Standardization [18] and are therefore elided here. This structure follows an existing example of a usability engineering project for a graphical user interface published in [20] (p. 29), however our interpretation differs in the mixed methods it employs.

Each of the four stages in the usability engineering process is outlined in Sects. 2.1 to 2.4 below. The heavy weighting given to formative evaluation in the standard is reflected by the emphasis given to Sect. 2.3.

2.1   User Research

In the first stage, three methods are employed for the user research: interviews, focus groups and questionnaires. Participants are clinicians from the project team and adult T1D subjects. This part of the process is used to define the user classes, user characteristics and use environment.

Table 1. PEPPER usability engineering process.

USER research	1. Prepare USE SPECIFICATION: focus groups, interviews, surveys
Analysis	2. Identify characteristics for SAFETY: TASK analysis, document review
	3. Identify HAZARDS and HAZARDOUS SITUATIONS
	4. Identify and describe HAZARD-RELATED USE SCENARIOS
	5. Select USE SCENARIOS for SUMMATIVE EVALUATION
Iterative design and FORMATIVE EVALUATION	6. Establish USER INTERFACE SPECIFICATION
	7. Establish USER INTERFACE (UI) EVALUATION plan
	Phases 1: Analytical study. Phase 2: Empirical lab study. Phase 3: Contextual study
	8. Perform UI design, implementation and FORMATIVE EVALUATION
SUMMATIVE	9. Perform SUMMATIVE EVALUATION: repeat Phase 2

2.2 Analysis

This stage comprises four steps, with two overall goals. First, to analyse the data gathered in user research stage in order to develop the user interface specification required for the next stage. Second, to anticipate unintended use of the system and to evaluate risk accordingly. The steps are:

- Identify characteristics for safety to determine tasks that users are expected to perform.
- Identify hazards and hazardous situations for each individual project, validate each one, and collate the results into a risk management file.
- Identify hazard-related use scenarios to determine tasks that users are not intended to perform.
- Select hazard-related use scenarios for summative evaluation according to probability of occurrence and severity of harm.

The design of the task model is informed both by the user research and relevant literature. Key tasks are verified with users and modelled in a suitable framework. The unintended task scenarios are collated by examining user manuals and public databases [14, 33], which give information about safety incidents and recalls.

2.3   Iterative Design and Formative Evaluation

This stage forms the heart of the process described in this paper. Once the user interface specification is designed. It is formatively evaluated and the interface is iteratively redeveloped. The stage consists of the following steps:

Establish User Interface Specification. Results of the task analysis are converted to a user requirements document.

Establish UI Evaluation Plan. Standard techniques for the formative usability evaluation can be grouped into three phases. Two involve users, either in controlled or natural settings, and the third does not involve users, but instead engages usability experts. The process defined here employs all three methods in separate phases, interleaved with design and redevelopment steps in accordance with the iterative development methodology.

Phase 1: Analytical Study. Three procedures were chosen for this phase: heuristic evaluation, keystroke-level model (KLM) [27] and competitive analysis. Conducting more than one procedure will ensure meaningful results given that heuristic evaluation is subjective whereas KLM is quantitative. All procedures were implemented by dual-domain experts: computer scientists familiar with both standard human interface evaluation methods and T1D. Such participants are aware of dangers such as hypoglycaemia and hyperglycaemia and have been shown to find a significantly higher percentage of usability errors than usability experts [13].

Phase 2: Empirical Laboratory Study. The purpose of this study was to measure the performance of the system with regard to the usability goals of simplicity, effectiveness, efficiency, and satisfaction. The system was tested in one-to-one sessions, each one lasting two hours. During each session, investigators gave participants the same series of 13 scenarios to perform as in the KLM evaluation. The scenarios were evaluated using quantitative metrics as displayed in Table 2. The participants' behaviour was audio recorded, and the interaction of their hands on the device was video recorded.

Table 2. Usability metrics

Measurement	Attribute
Percentage of tasks solved	Effectiveness
Percentage of users able to complete a given task	Effectiveness
Number of clicks/touches to solve task	Efficiency
Time taken to solve tasks	Efficiency
Number of errors per task	Simplicity
Type of error (e.g. precision error - missed target, response error – user clicks multiple times, affordance error – wrong icon or incorrect gesture, mode error etc.)	Simplicity

Phase 3: Empirical Contextual Study. The PEPPER situated study was closely based on an existing method [29]. It lasted four weeks, involved 15 participants and included auto-ethnography, an initial interview, a diary study and a contextual group meeting. The diary study formed the heart of this usability study. Its purpose was used to understand the day-to-day user experience with PEPPER over a period of several weeks. The goal was to see how context affects the use of the technology and also to understand which features may affect motivation, either positively or negatively. Participants were asked to make diary entries each time they used the PEPPER bolus advisor and they were also phoned at weekly intervals. The final step is an observational study of the group in a social setting at an informal location such as a cafe. The purpose is to validate the findings from the prior steps and to observe discourse about the experience.

Perform UI Design, Implementation and Formative Evaluation. Munzer's nested model [26] has been selected to conduct a user-centred design of visualisations of the patient's recorded data. This is preceded by an expert review of competitor data visualisation methods. Tufte's accessible complexity criteria [11] were adopted for this purpose. The results were used to inform the creation of multiple designs. These were evaluated in a participatory design workshop to create a single a wireframe for implementation in the first prototype. Once the interface was implemented, it undergoes the evaluation, according to the pre-defined plan.

2.4   Summative Evaluation

This final stage of the process comprises usability testing against the same usability goals, and using the same methods as in Phase 2 to ensure that the scores have improved. The risk management file is updated so that appropriate control measures exist to mitigate all hazard-related use scenarios.

3   Results

These methods were applied to the PEPPER handset. For reasons of space, we present limited preliminary results for the first two stages in Table 1: user research and analysis, focusing instead on the results of the iterative design and formative evaluation stage. The results of the last stage: summative evaluation, will be published once the study has concluded.

3.1   User Research

Two rounds of interviews were conducted, followed by three focus groups, where patients completed questionnaires. The end-users of the PEPPER system fall into four classes:

1. Patient (MDI);
2. Patient (CSII);

3. Clinician responsible for revision;
4. Clinician monitoring the personal health record.

The latter two classes may overlap, but the others do not. Patient inclusion and exclusion criteria were determined from the clinical protocol. In addition, system developers monitoring some aspects of the interface are another class of users, but this class is not included in the study.

   User characteristics were also recorded, not simply demographics but also susceptibilities of this population that might affect the ability to interact with the system, such as visual impairment. The use environment includes all contexts, such as storage, methods of conveyance, noise, and lighting conditions.

3.2   Analysis

Data analysed from the user research (focus groups, interviews and surveys) was used to create 13 use case scenarios for formative and summative evaluations, see Table 3. For the more complicated tasks, Unified Modelling Language (UML) descriptions were also developed.

Table 3. Scenarios

Number	Scenario
1	Open the application on the smartphone (MDI)/turn on the device (CSII)
2	Calibrate the continuous glucose meter using the Bluetooth capiliary meter
3	Locate and state the most recent bolus in the last 12 h
4	Locate and state the most recent carbohydrate intake in the last 12 h
5	Request bolus advice, for the following situation: (a) 45 g of carbohydrates (b) Medium meal absorption (c) Low intensity, non–aerobic past exercise (d) No planned exercise (e) No hormone cycle info (f) Low alcohol consumption (g) Not stressed, good mood
6	Accept or reject the bolus advice
7	Record 65 g of carbohydrates without requesting bolus advice
8	Record a 3 Insulin Unit bolus without requesting bolus advice
9	Locate and state the current blood glucose level
10	Locate and state the average blood glucose level for the past week
11	Locate and state the percentage of time have you spent in a 'high' blood glucose state over the past week
12	Locate and state the highest BG value in the last 12 h
13	Locate and state the total bolus over the last 12 h

Hazard analysis was documented and categorised according to severity and likelihood. Examples include water damage to the handset and loss of communication from sensors.

3.3   Iterative Design and Formative Evaluation

Elaborated results of this stage and corresponding steps are presented below. It includes the specification of user interface, the evaluation plan, and PEPPER design.

Establish User Interface Specification. The results of this step are embodied in the user requirements manual. The document was shared with the other project members.

Establish UI Evaluation Plan. This step comprises three phases: analytics study, empirical laboratory study and empirical contextual study.

Phase 1: Analytical Study. Includes (a) heuristic evaluation, (b) keystroke-level model (KLM), and (c) competitive analysis.

(a) Heuristic Evaluation. For the heuristic evaluation, three experts conducted the evaluation independently. In total 11 heuristics were used (Table 4). The heuristic with prefix Z was derived from Zhang et al. [35], heuristics with prefix B were derived from Bertini et al. [5], and heuristics with the prefix D were derived from Tufte [11] and Thimbleby [32]. Heuristics were specifically chosen in order to facilitate the analysis of medical devices, where there is a particular focus on safety, and to include more accessibility considerations specifically for people with diabetes susceptible to visual defects as a result of their condition.

Table 4. Heuristics

ID	Heuristics
Z6	Informative feedback
B1	Visibility of system status and losability/findability of the device
B2	Match between system and the real world
B3	Consistency and mapping
B4	Good ergonomics and minimalist design
B5	Ease of input, screen readability and glanceability
B6	Flexibility, efficiency of use and personalisation
B7	Aesthetic, privacy and social conventions
B8	Realistic error management
D1	Safe and efficient numerical data entry
D2	Clear and meaningful numerical data visualization

Each heuristic was given a score based on Nielsen's Severity Ranking Scale [28]. A summary of evaluators' results is shown in Table 5. Data were analysed both quantitatively and qualitatively. The former comprised computation of the mean and standard deviation of the severity of violation for each heuristic. The latter consisted of a thematic analysis of the comments provided by evaluators in relation to each sub-heuristic with mean violation above a given threshold. Results were presented to the evaluators at a debriefing session to develop recommendations on how improving the prototype to mitigate the problems that have been identified prior to user testing. The list of the recommendations is given in Appendix A.

Table 5. Pepper application heuristic evaluation results.

Heuristic	Severity Ranking (0–4)				
	Evaluator 1	Evaluator 2	Evaluator 3	Average	STDev
Z6	0.50	0.50	0.00	0.33	0.29
B1	2.00	1.00	2.50	1.83	0.76
B2	1.67	0.67	1.33	1.22	0.51
B3	0.00	0.00	1.00	0.33	0.58
B4	0.50	0.50	0.00	0.33	0.29
B5	0.40	0.40	0.80	0.53	0.23
B6	1.67	0.67	1.33	1.22	0.51
B7	1.33	0.67	1.67	1.22	0.51
B8	2.50	1.00	2.00	1.83	0.76
D1	1.75	2.00	1.50	1.75	0.25
D2	0.67	0.33	0.00	0.33	0.33

(b) Keystroke-Level Model (KLM). Keystroke level modelling (KLM) was used to examine a selected set of scenarios of the PEPPER system (Table 3). Since KLM was not designed for touch-screen devices it was adapted using a more modern variant (Touch level modelling (TLM) [30]) to evaluate the handsets.

For the evaluation, a single expert conducted the tasks to provide a baseline of ideal timings for comparison with the results of Phase 2.

The results of the handset evaluation are in Table 6. Tasks with zero touches are tasks that require viewing an item on the home screen. Task 2 and task 5 were classified as inefficient tasks due to the high number of touches whereas Task 10 and Task 11 were classified as unclear as it is not obvious to the user that they have to touch the filter icon. For Task 2, the expert recommended allowing manual calibration without using the Bluetooth and as for Task 5 the recommendation was to reduce the number of parameters and touches. The expert also suggested that the statistics screen should be updated automatically.

Table 6. TLM results.

Task	Time (sec)	Touches
1	2.11	3
2	14.42	9
3	2.40	0
4	2.40	0
5	11.36	30
6	0.70	2
7	3.85	7
8	3.66	6
9	2.40	0
10	4.80	6
11	4.80	6
12	2.40	0
13	3.60	0

(c) Competitive Analysis. For the competitive analysis, the following procedure was used to conduct an analysis of competitor products:

1. Identify competitors for handset;
2. Determine a key set of tasks that are possible for both PEPPER and competitors;
3. Perform heuristic evaluation of each system and compare results;
4. Execute the selected set of tasks on each system and record the interactions using the KLM/TLM method.

The Cellnovo handset [8] was identified as the most appropriate competitor to the PEPPER handset with a set of similar tasks, and therefore, an additional heuristic evaluation was conducted on Cellnovo using the same heuristics as the PEPPER evaluation (Table 4). The results are compared in Table 7. Both devices score highly for all heuristics. However, it can be seen that PEPPER scores worse than Cellnovo in heuristics B1, B4 and D1. These issues were therefore addressed in the final recommendations.

Finally, the KLM/TLM evaluation was conducted on both Cellnovo and PEPPER handsets. Comparison of summary scores are given in Table 8. Not all of the tasks in the model could be executed on the Cellnovo device, therefore the results are limited to those than could be done. It is clear from this that PEPPER scores better than Cellnovo in all but 2 tasks: 10 and 13. These tasks are therefore improved in the final recommendations.

Table 7. Heuristic evaluation of Cellnovo handset versus PEPPER handset (in sec).

Heuristic	Cellnovo Average	PEPPER
Z6	0.92	0.33
B1	1.75	1.83
B2	2.00	1.22
B3	1.13	0.33
B4	0.25	0.33
B5	1.30	0.53
B6	2.00	1.22
B7	1.50	1.22
B8	2.00	1.83
D1	1.19	1.75
D2	1.08	0.33

Table 8. Estimated time per task of Cellnovo handset versus PEPPER handset (in secs).

Task ID	1	3	4	5	9	10	12	13
PEPPER	2.11	2.40	2.40	11.36	2.40	4.80	2.40	3.60
Cellnovo	2.10	2.70	2.70	51.10	2.40	1.80	2.51	2.51

Phase 2: Empirical Laboratory Study. Fifteen patients were enrolled in the handset study: seven in Spain and eight in the UK. Four clinicians participated in the server study. Videos were analysed for each of the participants. Three out of the 15 files were corrupted so the results reported concerned the 12 remaining participants only using the metrics from Table 2.

- **Percentage of tasks solved.** The majority of tasks were solved by all of the users.
- **Percentage of users able to complete a given task.** The majority of users were able to solve all of the tasks.
- **Number of clicks/touches to solve task.** The tasks that stand out as requiring too much interaction are 2 and 5.
- **Time taken to solve tasks.** The tasks that stand out as taking too much time are 2 and 5.
- **Number of errors per task.** Significant errors occurred on Tasks 2, 7, and 13.
- **Type of error.** All error types have been recorded and documented in a spreadsheet. Most errors were navigational, but some were numeric. A more comprehensive analysis remains to be performed.

The SUS questionnaire was used to assess the users' satisfaction [6]. The SUS scores were excellent for the handset (74.3%) as seen in Fig. 3. Video data analysis showed there were few errors and most tasks were completed, showing high simplicity and effectiveness respectively. Inefficient tasks were identified from the average times. Think-aloud comments contributed to the recommendations for redesign.

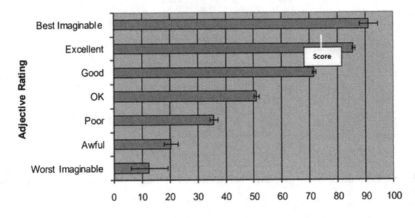

Fig. 3. SUS score.

Phase 3: Empirical Contextual Study. Early findings show that participants liked the system; more specifically, PEPPER encouraged them to be more diligent about logging their food diary and physical activity. The system also allowed them to visualise the repercussions of their diet on their glucose level. One participant reported that PEPPER made him/her aware of some incidents of hypoglycemia that they did not know of before the incidents happened during the night. Participants also reported trusting the recommendations and alarms. However, PEPPER makes them constantly aware of their condition, and concerned about glucose targets. They also thought that there were too many parameters required for the CBR model. These preliminary results have important implications for developers of AI self-management systems for diabetes and other conditions.

Perform UI Design, Implementation, and Formative Evaluation. Five paper prototypes were initially produced, influenced by existing devices and apps. Paper prototypes were then used to obtain feedback and suggestions for improvement, and to prompt discussion amongst participants in a participatory design workshop. Feedback was used to present the refined designs (Fig. 4).

3.4   Summative Evaluation

The final step of the usability engineering process is scheduled to be implemented during the year 2019. The results of the evaluation will be published.

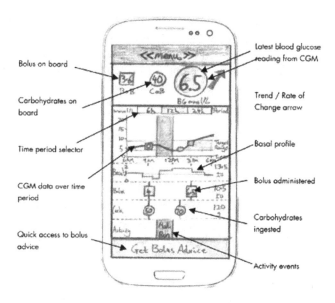

Fig. 4. Revised PEPPER dashboard.

4 Discussion

The research so far has suggested further research questions, including:

- How is data visualisation useful in developing trust and increasing acceptance of the bolus advice? For example, confidence might grow if a user can easily see that the safety system intervenes to recommend corrective action when one is moving towards hypoglycemia.
- How is adherence affected by context? For instance, one individual with experience of a similar system [7] reported removing the device during a music festival to avoid the data being viewed by the clinician. This points to privacy questions in the relationship between patient and clinician.

We also believe that there is a future need for domain-specific resources to assist with user-centred design of AI for diabetes decision support. For example, verification, validity and security are all research priorities for socially acceptable AI [31]. Therefore, there could be some merit in publishing a validated, definitive core task model, verified using formal techniques [2,23]. Similarly, empirically validated heuristics to support safety in medical mobile devices are lacking. For example, there is a trade-off between speed and safety; digit-based keypads are fast but may lead to larger mistakes than five-key interfaces. Standards and guidance exist to help prevent numerical errors, and these could inform such domain-specific heuristics.

Some of the user requirements are beyond the scope of this project. For example, one suggestion was to design an interface for manual input that adapts according to context and individual behaviour patterns, but this is not achievable in the timescale. There was also a demand for explanations to be provided for the bolus advice, as this might also engender trust in AI. This proposal is not feasible in the study however, since it would compromise the double-blind methodology.

5   Conclusion

In this paper we have given an overview of how our project is conducting a process that adheres to international standards for consideration of human factors in the design of a medical device. We have also proposed a method in which a situated study can be incorporated into standards, to fill a perceived gap around evaluation of systems in context. Finally, the results of the formative evaluation stage has been presented.

Acknowledgement. This work has received funding from the EU Horizon 2020 research and innovation programme under grant agreement No. 689810. We thank all partners of the PEPPER consortium, in particular clinical teams in UK and Spain who conducted the Phase 2 of the formative usability evaluation. Ethical approval has been obtained from the relevant authorities for all elements involving users.

Appendix A

The following section of this report document the notable and agreed issues and recommendations against each heuristic during a debrief with the evaluators. In some cases, issues listed by the evaluator for a particular heuristic have been relocated to a more suitable heuristic following the debrief.

Z6 - Informative Feedback
Average score: 0.33
Standard deviation: 0.29
Result: No usability issues
Issue: Activity monitor connection is not informative.
Recommendation: Provide feedback and information on the connection status of the activity monitor.

B1 - Visibility of System Status and Losability/Findability of the Device
Average score: 1.83
Standard deviation: 0.76
Result: Minor usability issue
Issue: Network, activity monitor and CGM status are not clearly indicated.
Recommendation: Include clear indications of connection status for network, activity monitor and CGM on the status bar.

B2 - Match Between System and Real World

Average score: 1.22

Standard deviation: 0.51

Result: Cosmetic problem only

B2.1 Issue: Ambient temperature location on the Get Bolus Advice interface is not intuitive.

B2.1 Recommendation: Relocate ambient temperature to the other category.

B2.2 Issue: The location of obtaining a capillary blood glucose reading is not intuitive.

B2.2 Recommendation: Include the ability to obtain a capillary blood glucose reading from the main menu and calibration interfaces.

B2.3 Issue: The application interface is always in portrait orientation.

B2.3 Recommendation: Appropriate landscape interfaces should be included. For example, changing the visualizations on the dashboard when in landscape orientation.

B2.4 Issue: The Android back button (bar at the bottom) interfaces with the interface due to full screen mode.

B2.4 Recommendation: Change PEPPER to a non-full screen application for MDI users on the handset or ensure that this bar does not interfere with the PEPPER interface.

B3 - Consistency and Mapping

Average score: 0.33

Standard deviation: 0.58

Result: No usability issues

Issue: The target blood glucose thresholds do not match what actually happens, stating that the lowest/highest possible value is invalid.

Recommendation: Investigate and correct the target blood glucose thresholds to ensure the valid range displayed to the user is correct or that valid inputs are accepted.

B4 - Good Ergonomics and Minimalist Design

Average score: 0.33

Standard deviation: 0.29

Result: No usability issues

Issue: The application has a mixture of light and dark interfaces.

Recommendation: Use a consistent colour palette throughout the interfaces.

B5 - Ease of Input, Screen Readability and Glanceability

Average score: 0.53

Standard deviation: 0.23

Result: Cosmetic problem only

B5.1 Issue: The abbreviation IU needs clarifying.

B5.1 Recommendation: Change the abbreviation IU to Units (based on clinician advice).

B5.2 Issue: The menu can only be accessed from the dashboard interface.

B5.2 Recommendation: Replace the home button with the menu option on all interfaces.

B5.3 Issue: The CGM cannot be automatically calibrated from the capillary blood glucose meter.

B5.3 Recommendation: Add a calibration option which obtains the blood glucose reading from the capillary blood glucose meter and automatically calibrates the CGM.

B6 - Flexibility, Efficiency of Use and Personalisation

Average score: 1.22
Standard deviation: 0.51
Result: Cosmetic problem only

B6.1 Issue: No ability to customise the application.

B6.1 Recommendation: Add some degree of customisation. For example, quick links and custom colour palettes.

B6.2 Issue: The keyboard overlaps input fields on the Get Bolus Advice interface.

B6.2 Recommendation: Remove the next option from the keyboard when inputting on the Get Bolus Advice interface, instead include the done button to close the keyboard.

B6.3 Issue: Some interfaces use number pickers rather than a keyboard for data entry. For consistency and speed these should all be keyboard inputs.

B6.3 Recommendation: Replace number pickers with appropriate keyboards, limiting the characters/digits to only valid inputs.

B7 - Aesthetic, Privacy and Social Conventions

Average score: 1.22
Standard deviation: 0.51
Result: Cosmetic problem only

B7.1 Issue: There are some harsh edges on the interface.

B7.1 Recommendation: Use the Android material design features to soften the edges, for example z-index.

B7.2 Issue: The applications lock screen does not provide security in its present state.

B7.2 Recommendation: The lock screen is not needed as the phone has its own lock screen. On the CSII version, the tap the Xs should be replaced by a personalised PIN to prevent unauthorized use.

B7.3 Issue: The application does not indicate that data has been transmitted successfully.

B7.3 Recommendation: Include an interface of data sent to the PEPPER server application, perhaps on the Events interface.

B8 - Realistic Error Management

Average score: 1.83
Standard deviation: 0.76
Result: Minor usability issues

Issue: The application does not provide any undo functionality for adding boluses/meals.

Recommendation: Include the ability to undo the previous bolus/meal input.

D1 - Safe and Efficient Numerical Data Entry

Average score: 1.75

Standard deviation: 0.25

Result: Minor usability issues

D1.1 Issue: Not all inputs are validated to ensure safe data entry.

D1.1 Recommendation: Clinically approved thresholds should be introduced for all inputs. For example, carbohydrates on the Get Bolus Advice interface.

D1.2 Issue: There is no ability to obtain a blood glucose reading from the capillary blood glucose meter on the Get Bolus Advice interface in the event of CGM failure.

D1.2 Recommendation: Add the option to obtain a capillary blood glucose reading via Bluetooth on the Get Bolus Advice interface. Performing this task should be mandatory if the application has not received data from the CGM recently.

D2 - Clear and Meaningful Numerical Data Visualization

Average score: 0.33

Standard deviation: 0.33

Result: No usability issues

D2.1 Issue: Some of the text is small and difficult to read.

D2.1 Recommendation: Introduce a minimum text size to ensure the information is clear. Additionally display more information when clicking on the carbohydrate/bolus/activity on the dashboard with a large text size and precise underlying data.

D2.2 Issue: Bolus and carbohydrate values on the dashboard can overlap if input within a small time interval.

D2.2 Recommendation: Merge overlapping data on the dashboard visualization (carbs and bolus) and indicate that this is multiple inputs and provide the user the ability to tap to view details.

References

1. Aamodt, A., Plaza, E.: Case-based reasoning: foundational issues, methodological variations, and system approaches. AI Commun. **7**(1), 39–59 (1994)
2. Abrial, J.R., Abrial, J.R.: The B-book: Assigning Programs to Meanings. Cambridge University Press, Cambridge (2005)
3. Årsand, E., et al.: Mobile health applications to assist patients with diabetes: lessons learned and design implications. J. Diab. Sci. Technol. **6**(5), 1197–1206 (2012)
4. Bellazzi, R.: Telemedicine and diabetes management: current challenges and future research directions. J. Diab. Sci. Technol. **2**(1), 98–104 (2008)

5. Bertini, E., Catarci, T., Dix, A., Gabrielli, S., Kimani, S., Santucci, G.: Appropriating heuristic evaluation methods for mobile computing. In: Handbook of Research on User Interface Design and Evaluation for Mobile Technology, pp. 780–801. IGI Global (2008)

6. Brooke, J., et al.: SUS-A quick and dirty usability scale. Usability Eval. Ind. **189**(194), 4–7 (1996)

7. Brown, D.: Reflections on the role and design of visualizations to support patients with type 1 diabetes (2016, in preparation)

8. Cefai, J.J., Shapley, J.L.: Micro-valve. US Patent 8,048,041, 1 November 2011

9. Cellnovo: Cellnovo system (2007). https://www.cellnovo.com/en/homepage

10. Davis, F.D.: Perceived usefulness, perceived ease of use, and user acceptance of information technology. MIS Q. **13**, 319–340 (1989)

11. Tufte, E.: The Visual Display of Quantitative Information. Graphics Press, Cheshire (2001). vol. 4, no. 5, p. 6

12. Fast, E., Horvitz, E.: Long-term trends in the public perception of artificial intelligence. In: Proceedings of the Thirty-First AAAI Conference on Artificial Intelligence, pp. 963–969. International Joint Conferences on Artificial Intelligence, Menlo Park, Calif (2017)

13. Georgsson, M., Staggers, N., Weir, C.: A modified user-oriented heuristic evaluation of a mobile health system for diabetes self-management support. Comput. Inf. Nurs. **34**(2), 77 (2016)

14. GOV.UK: Medicines and healthcare products regulatory agency. Alerts and recalls for drugs and medical devices (2016). https://www.gov.uk/drug-device-alerts

15. Hengstler, M., Enkel, E., Duelli, S.: Applied artificial intelligence and trust–the case of autonomous vehicles and medical assistance devices. Technol. Forecast. Soc. Chang. **105**, 105–120 (2016)

16. Herrero, P., López, B., Martin, C.: PEPPER: patient empowerment through predictive personalised decision support. In: ECAI Workshop on Artificial Intelligence for Diabetes, pp. 8–9 (2016)

17. Herrero, P., et al.: Safety layer for an insulin delivery system. In: Abstracts from ATTD 2017 10th International Conference on Advanced Technologies & Treatments for Diabetes Paris, France, 15-18 February 2017, p. A-62 (2017). https://doi.org/10.1089/dia.2017.2525

18. IEC: International electrotechnical commission (2018). http://www.electropedia.org/

19. International Standard IEC: 62366-1: 2015 medical devices part 1: application of usability engineering to medical devices. International Organization for Standardization, Geneva (2015)

20. International Standard IEC: 62366-2: 2016 medical devices part 2: Guidance on the application of usability engineering to medical devices. International Organization for Standardization, Geneva (2016)

21. ISMP: Institute for safe medication practices, fluorouracil incident root cause analysis report (2007). https://www.ismp-canada.org

22. ISO: International organization for standardization online browsing platform (2018). https://www.iso.org/obp/ui/

23. Kirwan, B., Ainsworth, L.K.: A Guide to Task Analysis: The Task Analysis Working Group. CRC Press, Boca Raton (1992)

24. Klonof, D.C.: The current status of bolus calculator decision-support software. J. Diab. Sci. Technol. **6**(5), 990–994 (2012)

25. Klonoff, D.C.: Telemedicine for diabetes: current and future trends. J. Diab. Sci. Technol. **10**(1), 3–5 (2016)

26. Munzner, T.: A nested process model for visualization design and validation. IEEE Trans. Vis. Comput. Graph. **6**, 921–928 (2009)

27. Nielsen, J.: Usability Engineering. Morgan Kauffmann, San Francisco (1994)

28. Nielsen, J.: Severity ratings for usability problems. Pap. Essays **54**, 1–2 (1995)

29. O'Kane, A.: Individual differences and contextual factors influence the experience and practice of self-care with type 1 diabetes technologies. Ph.D. thesis, UCL (University College London) (2016)

30. Rice, A.D., Lartigue, J.W.: Touch-level model (TLM): evolving KLM-GOMS for touchscreen and mobile devices. In: Proceedings of the 2014 ACM Southeast Regional Conference, p. 53. ACM (2014)

31. Russell, S., Dewey, D., Tegmark, M.: Research priorities for robust and beneficial artificial intelligence. AI Mag. **36**(4), 105–114 (2015)

32. Thimbleby, H.: Feature ignorance of interaction programming is killing people. Interactions **15**(5), 52–57 (2008)

33. USFDA: MAUDE - manufacturer and user facility device experience (2018). http://www.accessdata.fda.gov/scripts/cdrh/cfdocs/cfMAUDE/TextSearch.cfm

34. USFDA, et al.: Applying human factors and usability engineering to optimize medical device design. FDA Center for Devices and Radiological Health: Silver Spring, MD, USA (2011)

35. Zhang, J., Johnson, T.R., Patel, V.L., Paige, D.L., Kubose, T.: Using usability heuristics to evaluate patient safety of medical devices. J. Biomed. Inform. **36**(1–2), 23–30 (2003)

Automated Pain Detection in Facial Videos of Children Using Human-Assisted Transfer Learning

Xiaojing Xu[1]([✉]) [iD], Kenneth D. Craig[2] [iD], Damaris Diaz[3] [iD],
Matthew S. Goodwin[4] [iD], Murat Akcakaya[5] [iD], Büşra Tuğçe Susam[5] [iD],
Jeannie S. Huang[3] [iD], and Virginia R. de Sa[6] [iD]

[1] Department of Electrical and Computer Engineering,
University of California San Diego, La Jolla, CA, USA
xix068@ucsd.edu
[2] Department of Psychology, University of British Columbia, Vancouver, BC, Canada
kcraig@psych.ubc.ca
[3] Rady Childrens Hospital and Department of Pediatrics,
University of California San Diego, La Jolla, CA, USA
{dad003,jshuang}@ucsd.edu
[4] Department of Health Sciences, Northeastern University, Boston, MA, USA
m.goodwin@northeastern.edu
[5] Department of Electrical and Computer Engineering, University of Pittsburgh,
Pittsburgh, PA, USA
akcakaya@pitt.edu,tugcebusraiu@gmail.com
[6] Department of Cognitive Science, University of California San Diego,
La Jolla, CA, USA
desa@ucsd.edu

Abstract. Accurately determining pain levels in children is difficult, even for trained professionals and parents. Facial activity provides sensitive and specific information about pain, and computer vision algorithms have been developed to automatically detect Facial Action Units (AUs) defined by the Facial Action Coding System (FACS). Our prior work utilized information from computer vision, i.e., automatically detected facial AUs, to develop classifiers to distinguish between pain and no-pain conditions. However, application of pain/no-pain classifiers based on automated AU codings across different environmental domains results in diminished performance. In contrast, classifiers based on manually coded AUs demonstrate reduced environmentally-based variability in performance. In this paper, we train a machine learning model to recognize pain using AUs coded by a computer vision system embedded in a software package called iMotions. We also study the relationship between iMotions (automatically) and human (manually) coded AUs. We find that AUs coded automatically are different from those coded by a human trained in the FACS system, and that the human coder is less sensitive to environmental changes. To improve classification performance in the current work, we applied transfer learning by training another machine learning model to map automated AU codings to a subspace of manual

© Springer Nature Switzerland AG 2019
F. Koch et al. (Eds.): AIH 2018, LNAI 11326, pp. 162–180, 2019.
https://doi.org/10.1007/978-3-030-12738-1_12

AU codings to enable more robust pain recognition performance when only automatically coded AUs are available for the test data. With this transfer learning method, we improved the Area Under the ROC Curve (AUC) on independent data from new participants in our target domain from 0.67 to 0.72.

Keywords: Automated pain detection · Transfer learning · Facial action units · FACS

1 Introduction

In the classic model of machine learning, scientists train models on a collected dataset to accurately predict a desired outcome and then apply learned models to new data measured under identical circumstances to validate performance. Given the notable variation in real world data, it is tempting to apply learned models to data collected under similar but non-identical circumstances. However, performance in such circumstances often deteriorates due to unmeasured factors not accounted for between the original and new datasets. Nevertheless, knowledge can be extracted in these scenarios. Transfer learning, or inductive transfer in machine learning parlance, focuses on using knowledge gained from solving one problem to improve performance on a different but related problem [1]. The present paper describes application of transfer learning to the important clinical problem of automated pain detection in children.

Accurate measurement of pain severity in children is difficult, even for trained professionals and parents. This is a critical problem as over-medication can result in adverse side-effects, including opioid addiction, and under-medication can lead to unnecessary suffering [2].

The current clinical gold standard and most widely employed method of assessing clinical pain is patient self-report [3]. However, this subjective method is vulnerable to self-presentation bias. Consequently, clinicians often distrust pain self-reports, and find them more useful for comparisons over time within individuals, rather than comparisons between individuals [4]. Further, infants, young children, and others with communication/neurological disabilities do not have the ability or capacity to self-report pain levels [3,5,6]. As a result, to evaluate pain in populations with communication limitations, observational tools based on nonverbal indicators associated with pain have been developed [7].

Of the various modalities of nonverbal expression (e.g., bodily movement, vocal qualities of speech), it has been suggested that facial activity provides the most sensitive, specific, and accessible information about the presence, nature, and severity of pain across the life span, from infancy [8] to advanced age [9]. Moreover, observers largely consider facial activity during painful events to be a relatively spontaneous reaction [7].

Evaluation of pain based on facial indicators requires two steps: (1) Extraction of facial pain features and (2) pain recognition based on these features. For step (1), researchers have searched for reliable facial indicators of pain, such as anatomically-based, objectively coded Facial Action Units (AUs) defined by the Facial Action Coding System (FACS) [10, 11]. (Visualizations of facial activation units can also be found at https://imotions.com/blog/facial-action-coding-system/). However, identifying AUs traditionally requires time intensive offline coding by trained human coders, limiting application in real-time clinical settings. Recently, algorithms to automatically detect AUs [11] have been developed and implemented in software such as iMotions (imotions.com) allowing automatic output of AU probabilities in real-time based on direct recording of face video. In step (2), machine learning algorithms such as linear models [5], SVM [12], and Neural Networks [13] have been used to automatically recognize pain based on facial features.

Although a simple machine learning model based on features extracted by a well-designed algorithm can perform well when training and test data have similar statistical properties, problems arise when the data follow different distributions, as happens, for example, when videos are recorded in two different environments. We discovered this issue when training videos were recorded in an outpatient setting and test videos in the hospital. One way to deal with this problem is to use transfer learning, which discovers "common knowledge" across domains and uses this knowledge to complete tasks in a new domain with a model learned in the old domain [14]. In this work, we show that features extracted from human-coded (manual) AUs are less sensitive to domain changes than features extracted from iMotions (automated) AU codings, and thus develop a simple method that learns a projection from automated features onto a subspace of manual features. Once this mapping is learned, future automatically coded data can be transformed to a representation that is more robust between domains. In this work, we use a neural network model to learn a mapping from automated features to manual features, and another neural network model to recognize pain using the mapped facial features.

To summarize, our contributions of this work include demonstrating that:

- Manually/automatically coded AUs can be used to successfully recognize clinical pain in videos with machine learning.
- Environmental factors modulate the ability of automatically coded AUs to recognize clinical pain in videos.
- Manually coded AUs (especially previously established "pain-related" ones) can be used to successfully recognize pain in videos with machine learning across different environmental domains.
- Automatically coded AUs from iMotions do not directly represent or correlate with AUs defined in FACS.
- Transfering automated features to the manual feature space improves automatic recognition of clinical pain across different environmental domains.

This work was presented at the Joint Workshop on Artificial Intelligence in Health and a shorter version of this paper appeared in the proceedings [15].

2   Methods

2.1   Participants

One hundred and forty-three pediatric research participants (94 males, 49 females) aged 12 [10, 15] (median [25%, 75%]) years old and primarily Hispanic (78%) who had undergone medically necessary laparoscopic appendectomy were videotaped for facial expressions during surgical recovery. Videos were subsequently categorized into two conditions: pain and no-pain. Participating children had been hospitalized following surgery for post-surgical recovery and were recruited for participation within 24 h of surgery at a pediatric tertiary care center. Exclusion criteria included regular opioid use within the past six months, documented mental or neurological deficits preventing study protocol compliance, and any facial anomaly that might alter computer vision facial expression analysis. Parents provided written informed consent and youth gave written assent [16]. The local institutional review board approved the research protocol.

Table 1. Numbers of samples at different pain levels in each visit.

Pain level	0	1	2	3	4	5	6	7	8	9	10
V1		16	12	18	28	31	26	26	19	24	15	11
V2		4	18	24	40	21	23	16	13	14	8	4
V3	166	17	3	1	0	0	0	0	0	0	0

2.2   Experimental Design and Data Collection

Data were collected over three visits (V): V1 within 24 h after appendectomy; V2 within the calendar day after the first visit; and V3 at a follow-up visit 25 [19, 28] (median [25%, 75%]) days postoperatively when pain was expected to have fully subsided. Data were collected in two environmental conditions: V1 and V2 in hospital and V3 in the outpatient setting. At every visit, two 10-second videos (60 frames per second at 853 × 480 pixel resolution) of the face were recorded while manual pressure was exerted at the surgical site for 10 seconds (equivalent of a clinical examination). During hospital visits (V1, V2), participants were lying in the hospital bed with the head of the bed raised. In the outpatient lab in V3, they were seated in a reclined chair. Participants rated their pain level during manual pressure using a 0–10 Numerical Rating Scale, where 0 = no-pain and 10 = worst pain ever. For classification purposes, and following convention used by clinicians for rating clinically significant pain [17], videos with pain ratings of 0–3 were labeled as no-pain, and videos with pain ratings of 4–10 were labeled as pain. Two hundred and fifty-one pain videos were collected from V1/2, 160 no-pain videos were collected from V1/2, and 187 no-pain videos were collected from V3. The numbers of samples collected for different pain levels and visits are shown in Table 1. Note that all V3 data are labeled as no-pain and there are only 4 pain ratings over 1 in V3. In contrast, the majority of no-pain data in V1 and V2 are ratings of 2 and 3. Figure 1 "All Data" demonstrates the distribution of pain and no-pain videos across environmental conditions.

	Visit 1 and Visit 2 (in hospital)		Visit 3 (in outpatient lab)
All Data	Pain	No Pain	
Data Domain 1 (D1)	Pain	No Pain	
Data Domain 2 (D2)	Pain		No Pain

Fig. 1. Data domain illustration. The area of category is not proportional to the number of samples.

AU	FACS name	AU	FACS name
1	Inner brow raiser	15	Lip corner depressor
2	Outer brow raiser	17	Chin raiser
4	Brow lowerer	18	Lip pucker
5	Upper lid raiser	20	Lip stretcher
6	Cheek raiser and Lid compressor	23	Lip tightener
7	Lid tightener	24	Lip pressor
9	Nose wrinkler	25	Lips part
10	Upper lip raiser	26	Jaw drop
12	Lip corner puller	28	Lip suck
14	Dimpler	43	Eyes closed

Fig. 2. FACS names (descriptions) of 20 AUs coded by iMotions. AUs 1–7 and 43 are upper face AUs, and the others are lower face AUs.

2.3   Feature Extraction

For each 10-second video sample we extracted AU codings per frame to obtain a sequence of AUs. This was done both automatically by iMotions software (www.imotions.com) and manually by a FACS trained human in a limited subset. A second trained human independently coded a subset of the videos coded by the first human. We then extracted features from the sequence of AUs.

Automated Facial Action Unit Detection: The iMotions software integrates Emotient's FACET technology (www.imotions.com/emotient), formally known as CERT [18]. In the described work, iMotions software was used to process videos to automatically extract 20 AUs as listed in Fig. 2 and three head pose indicators (yaw, pitch and roll) from each frame. The values of these codings represent estimated log probabilities of AUs, ranging from −4 to 4.

Manual Facial Action Unit Detection: A trained human FACS AU coder manually coded 64 AUs (AU1-64) for each frame of a subset (54%) of videos and labeled AU intensities (0–5, 0 = absence). In order to evaluate the reliability of the manual codings, we had another trained human coder code a subset (15%) of videos coded by the first human.

Feature Dimension Reduction: The number of frames in our videos was too large to use full sequences of frame-coded AUs. To reduce dimensionality, we applied 11 statistics (mean, max, min, standard deviation, 95th, 85th, 75th, 50th, 25th percentiles, half-rectified mean, and max-min) to each AU over all frames as in [5] to obtain 11×23 features for automatically coded AUs, and 11×64 features for manually coded AUs. We call these automated features and manual features, respectively. The range of each feature was rescaled to $[0, 1]$ to normalize features over the training data.

2.4   Machine Learning Models

Neural Network Model to Recognize Pain with Extracted Features: A neural network with one hidden layer was used to recognize pain with extracted automated or manual features. The number of neurons in the hidden layer was twice the number of neurons in the input layer, and the Sigmoid activation function $\sigma(x) = 1/(1 + \exp(-x))$ was used with batch normalization for the hidden layer. The output layer used Softmax activation and cross-entropy error.

Neural Network Model to Predict Manual Features with Automated Features: A neural network with the same structure was used to predict manual features from automated features, except that the output layer was linear and mean squared error was used as the loss function.

Model Training and Testing: Experiments were conducted in a participant-based (each participant restricted to one fold) 10-fold cross-validation fashion. Participants were divided into 10 folds, and each time 1 fold was used as the test set, and the other 9 folds together were used as the training set. We balanced classes for each participant in each training set by randomly duplicating samples from the under-represented class. One out of nine participants in the training sets were picked randomly as a nested-validation set for early stopping in the neural network training. A batch size of 1/8 the size of training set was used.

We then examined the receiver operating characteristic curve (ROC curve) which plots True Positive Rate against False Positive Rate as the discrimination threshold varies. We used Area under the Curve (AUC) to evaluate classification performance. We considered data from three domains (D) as shown in Fig. 1: (1) D1 with pain and no-pain both from V1/2 in hospital; (2) D2 with pain from V1/2 in hospital and no-pain from V3 from outpatient lab; and (3) All data, i.e., pain from V1/2 and no-pain from V1/2/3. The clinical goal was to be able to discriminate pain levels in the hospital; thus evaluation on D1 (where all samples were from the hospital bed) was the most clinically relevant evaluation.

Table 2. AUC for classification with SEM (standard error of the mean).

Train on	Test on	Automated	Manual	Automated "pain" features	Manual "pain" features
All	D1	0.61 ± 0.006	0.66 ± 0.006	0.63 ± 0.007	**0.69 ± 0.006**
D1	D1	0.58 ± 0.014	0.62 ± 0.008	0.61 ± 0.008	**0.65 ± 0.008**
D2	D1	0.57 ± 0.005	0.67 ± 0.007	0.62 ± 0.004	**0.7 ± 0.006**
All	D2	0.9 ± 0.005	0.79 ± 0.007	0.88 ± 0.005	0.8 ± 0.003
D1	D2	0.69 ± 0.011	0.68 ± 0.008	0.73 ± 0.012	0.73 ± 0.01
D2	D2	0.92 ± 0.01	0.79 ± 0.009	0.9 ± 0.007	0.8 ± 0.005

3 Analysis and Discussion

Data from 73 participants labeled by both human and iMotions were used through Sects. 3.1 to 3.5, and data from the remaining 70 participants using only automated (iMotions) AU codings were included for independent test set evaluation in the results section.

3.1 Automated Classifier Performance Varies by Environment

Using automated features, we first combined all visit data and trained a classifier to distinguish pain from no-pain. This classifier performed well in general (AUC $= 0.77 \pm 0.011$ on All data), but when we looked at different domains, the performance of D1 (the most clinically relevant in-hospital environment) was inferior to that on D2, as shown in data rows 1 and 4 under the "Automated" column in Table 2.

There were two main differences between D1 and D2, i.e., between V1/2 and V3 no-pain samples. The first was that in V1/2, participants still had some pain and their self-ratings were greater than 0, while in V3, no-pain ratings were usually 0 reflecting a "purer" no-pain signal. The second difference was that V1/2 occurred in the hospital with patients in beds and V3 videos were recorded in an outpatient setting with the participant sitting in a reclined chair. Lighting was also inherently different between hospital and outpatient environments. Since automated recognition of AUs is known to be sensitive to facial pose and lighting differences, we hypothesized that added discrepancy in classification performance between D1 and D2 was mainly due to the model classifying on environmental differences between V1/2 and V3. In other words, when trained and tested on D2, the classifier might distinguish "lying in hospital bed" vs "more upright in outpatient chair" as much as pain vs no-pain (this is similar to a computer vision algorithm doing well at recognizing cows by recognizing a green background).

In order to investigate this hypothesis and attempt to improve classification on the clinically relevant D1, we trained a classifier using only videos from D1. Within the "Automated" column, row 2 in Table 2 shows that performance on automated D1 classification does not drop much when D2 samples are removed from the training set. At the same time, training using only D2 data results in the worst classification on D1 (row 3), but the best classification on D2 (last row) as the network is able to exploit environmental differences (no-pain+more upright from V3, pain+lying-down from V1/2).

Figure 3(b) (LEFT) shows ROC curves of within and across domain tests for models trained on automated features in D2. The dotted (red) curve corresponds to testing on D2 (within domain) and the solid (blue) curve corresponds to testing on D1 (across domain). The model performed well on within domain classification, but failed on across domain tasks.

3.2   Classification Based on Manual AUs Are Less Sensitive to Environmental Changes

We also trained a classifier on manual AUs labeled by a human coder. Interestingly, results from the classifier trained on manual AUs showed less of a difference in AUCs between domains, with a higher AUC for D1 and a lower AUC for D2 relative to those with automated AUs (see Table 2 "Manual" and "Automated" columns). Overall, manual AUs appeared to be less sensitive to changes in the environment, reflecting the ability of human labelers to consistently code AUs without being affected by lighting and pose variations.

When we restricted training data from All to only D1 or only D2 data, classification performance using manual AUs went down, likely due to the reduction in training data, and training with D2 always gave better performance than training with D1 on both D1 and D2 test data, which should be the case since pain and no-pain samples in D2 are more discrepant in average pain rating. These results appear consistent with our hypothesis that human coding of AUs is not as sensitive as machine coding of AUs to environmental differences between V1/2 and V3.

Figure 3(b) (MIDDLE) displays ROC curves for manual features. As discussed above, in contrast to the plot on the left for automated features, manual coding performance outperformed automated coding performance in the clinically relevant test in D1. The dotted (red) curve representing within-domain performance is only slightly higher than the solid (blue) curve, likely due in part to the quality difference in no-pain samples in V1/2 and V3, and also possibly any small amount of environmental information that the human labeler was affected by. Note that ignoring the correlated environmental information in D2 (i.e., pain faces were more reclined and no-pain faces were more upright) resulted in a lower numerical performance on D2 but does not likely reflect worse classification of pain but instead the failure to "cheat" by using features affected by pose angle to classify all upright faces as "no-pain."

3.3   Restricting Manual AUs to Those Associated with Pain Improves Classification

In an attempt to reduce the influence of environmental conditions to further improve performance on D1, we restricted the classifier to the eight AUs

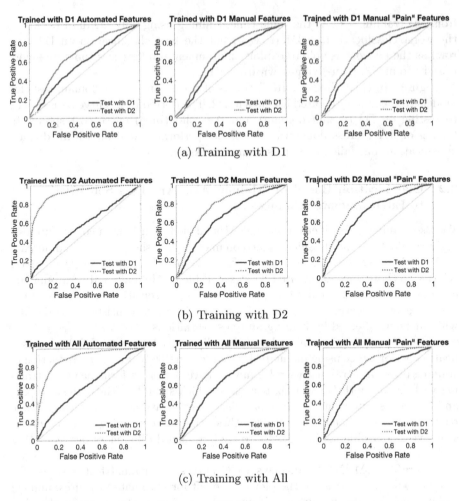

(a) Training with D1

(b) Training with D2

(c) Training with All

Fig. 3. ROC curves for classification on D1 and D2 using automated features (left), manual features (middle) and pain-related manual features (right), when the model is trained on (a) D1, (b) D2 and (c) All data. The dotted (red) lines are ROCs when the machine is able to use environment information to differentiate pain and no-pain conditions, and the solid (blue) lines show the machine's ability to discriminate between pain and no-pain based on AU information alone. The straight (yellow) line graphs the performance of random chance. (Color figure online)

consistently associated with pain: 4 (Brow Lowerer), 6 (Cheek Raiser), 7 (Lid Tightener), 9 (Nose Wrinkler), 10 (Upper Lip Raiser), 12 (Lip Corner Puller), 20 (Lip Stretcher), and 43 (Eyes Closed) [19, 20] as illustrated in Fig. 4 to obtain 11 (statistics) ×8 (AUs) features. Pain prediction results using these "pain" features are shown in the last two columns in Table 2. Results show that using only pain-related AUs improved classification performance of manual features. However, it did not seem to help as much for automated features.

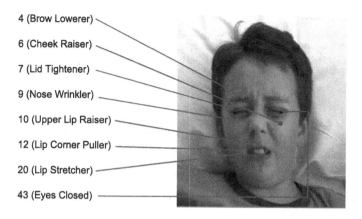

4 (Brow Lowerer)

6 (Cheek Raiser)

7 (Lid Tightener)

9 (Nose Wrinkler)

10 (Upper Lip Raiser)

12 (Lip Corner Puller)

20 (Lip Stretcher)

43 (Eyes Closed)

Fig. 4. Illustration of eight "pain-related" facial AUs.

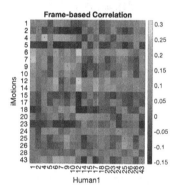

Fig. 5. Correlation matrix of AU pairs from automated and manual codings using All data.

Similarly, Fig. 3(b) (RIGHT) shows that limiting manual features to use only pain-related AUs further improved D1 performance when training with D2. We also employed PCA on pain-related features and found that performance in the hospital domain was similar if using four or more principal components.

In Fig. 3(a) and (c) we show ROC curves similar to Fig. 3(b) except with different training data. These curves correspond to row 2 and 5 (a), or 1 and 4 (c), under "Automated," "Manual," and "Manual 'Pain' Features" in Table 2.

3.4   iMotions AUs Are Different Than Manual FACS AUs

Computer Vision AU automatic detection algorithms have been programmed/trained on manual FACS data. However, we demonstrate differential performance of AUs encoded automatically versus manually. To understand the relationship between automatically encoded v. manually coded AUs, we computed correlations between binarized automatically coded AUs and manually

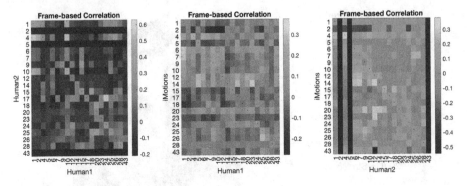

Fig. 6. Correlations of AU pairs from two of (1) iMotions; (2) human 1; and (3) human 2 on a subset of the data.

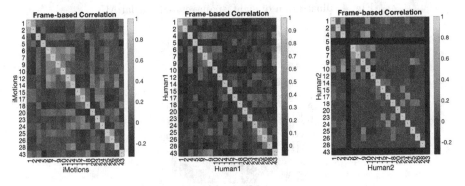

Fig. 7. Self-correlation matrices of AU pairs from iMotions or humans.

coded AUs at the frame level as depicted in Fig. 5. The FACS names corresponding to AU numbers are listed in Fig. 2, in which AUs 1, 2, 4, 5, 6, 7, 43 are upper face AUs and all others are lower face AUs. If two sets of AUs were identical, the diagonal of the matrix (marked with small centered dots) should yield the highest correlations, which was not the case. For example, manual AU 6 was highly correlated with automated AU 12 and 14, but had relatively low correlation with automated AU 6.

The correlation matrix shows that not only is our first human coder less affected by environmental changes, the AUs she coded are not in agreement with the automated AUs. Our second trained human coder (human 2) shows a better correlation with the coding of human 1 than between each human and iMotions, shown in Fig. 6 (LEFT). The correlation between each of the humans and the software on the same subset is shown in Fig. 6 (MIDDLE, RIGHT). This likely explains the reduced improvement by restricting the automated features model to "pain-related AUs" as these have been determined based on human FACS coded AUs (Fig. 8).

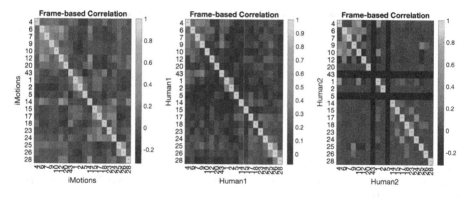

Fig. 8. Self-correlation matrices of AU pairs from iMotions or humans with "pain" AUs arranged together at the top left corner.

Table 3. AUC (and SEM) with transferred automated features.

Train on	Test on	All features	"Pain" features	7 PCs	4 PCs	1 PC
All	D1	0.61 ± 0.009	0.63 ± 0.009	0.68 ± 0.006	$\mathbf{0.69 \pm 0.008}$	0.65 ± 0.009
D1	D1	0.62 ± 0.009	0.64 ± 0.014	0.66 ± 0.012	$\mathbf{0.67 \pm 0.011}$	0.65 ± 0.009
D2	D1	0.58 ± 0.011	0.59 ± 0.01	0.66 ± 0.008	$\mathbf{0.68 \pm 0.006}$	0.66 ± 0.009
All	D2	0.82 ± 0.009	0.82 ± 0.009	0.76 ± 0.009	0.75 ± 0.012	0.7 ± 0.01
D1	D2	0.69 ± 0.009	0.71 ± 0.013	0.7 ± 0.015	0.71 ± 0.015	0.69 ± 0.011
D2	D2	0.88 ± 0.011	0.86 ± 0.006	0.76 ± 0.013	0.74 ± 0.01	0.7 ± 0.009

The self-correlation matrices between AUs in iMotions and the human coder are shown in Fig. 7. AUs coded by iMotions show higher correlations (between different iMotions coded AUs) than AUs coded by humans. Some human AU codings were also correlated, which is expected since specific AUs often occur together (e.g., AU 1 and 2 for inner and outer brow raiser and AU 25 and 26 for lips part and jaw drop) and other AUs tend to occur together in pain. This latter correlation of pain AUs is more evident in Fig. 4 which shows the same content as Fig. 7 except that in Fig. 4 the eight pain-related AUs are put together at the upper left corner to highlight their higher correlations. Interestingly, higher correlations within the pain AUs for iMotions coding was observed but the pattern is different.

3.5 Transfer Learning via Mapping to Manual Features Improves Performance

We have shown that manual codings are not as sensitive to domain change. However, manual coding of AUs is very time-consuming and not amenable to an automated real-time system. In an attempt to leverage manual coding to achieve similar robustness with automatic AUs, we utilized transfer learning and mapped automated features to the space of manual features. Specifically,

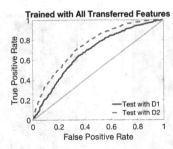
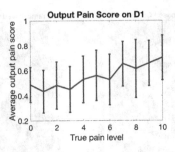

Fig. 9. ROC curves for classification on two domains using our transfer learning model (left) and plot of average model output pain score (with error bars indicating standard deviation) over true pain level (right).

we trained a neural network model to estimate manual features from automated features using data coded by both iMotions and a human. Separate models were trained to predict: manual features of 64 AUs, manual features of the eight pain-related AUs, and principal components (PCs) of the manual features of the eight pain-related AUs. PCA dimensionality reduction was used due to insufficient data for learning an accurate mapping from all automated AUs to all manual AUs.

Once the mapping network was trained, we used it to transform the automated features and trained a new network on these transformed data for pain/no-pain classification. The 10-fold cross-validation was done consistently so that the same training data was used to train the mapping network and the pain-classification network.

In Table 3, we show classification AUCs when the classification model was trained and tested with outputs from the prediction network. We observed that when using All data to train (which performed best), with the transfer learning prediction network, automated features performed much better in classification on D1 (0.68–0.69 compared to 0.61–0.63 in Table 2). Predicting four principal components of manual pain-related features yielded the best performance in our data. Overall, the prediction network helped in domain adaptation of a pain recognition model using automatically extracted AUs.

Figure 9 (LEFT) plots the ROC curves on two domains using the transfer learning classifier trained and tested using four predicted features. The model performed well in across-domain classification. Compared to Fig. 3(c) (LEFT), the transferred automated features showed properties more similar to manual features (Fig. 3(c) (RIGHT)), with smaller differences between performance on the two domains and higher AUC on the clinically relevant D1. Table 3 shows numerically how transfer learning helped automated features ignore environmental information in D2 like humans, and learn pure pain information that can be used in classification on D1.

Within-domain classification performance for D1 was also improved with the prediction network. These results show that by mapping to the manual feature space, automated features can be promoted to perform better in pain classification.

Figure 9 (RIGHT) plots output pain scores of our model tested on D1 versus 0–10 self-reported pain levels. The model output pain score increases with true pain level, indicating that our model indeed reflects pain levels.

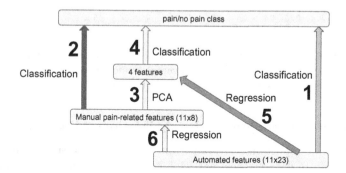

Fig. 10. Illustration of machine learning models. 1/2 are classifications using automated/manual pain features, in which 2 does better than 1. 3–4 can be done to reduce feature dimensions while maintaining performance. 6–2 and 5–4 are our transfer learning models, training a regression network to map automated features to a subspace of manual pain features before classification.

4   Results

In the previous section we showed that in Fig. 10 classification with pain-related pain features (2) performed better than automated features (1) on D1, which was the clinically relevant classification. We also found that applying PCA to manual features (3–4) does not change performance on D1 much. Thus, we introduced a transfer learning model to map automated features first to manual pain-related features (or the top few principal components of them), and then used the transferred features for classification (6–2 or 5–4). We obtained similar results to manual features on D1 with the transfer learning model (5–4) mapping to four principal components of manual features.

Table 2 shows that without our transfer learning method, training on all data and restricting to pain-related AUs results in the best performance using automated features for D1. And cross-validation results in Table 3 shows that with our method, using all data and predicting four PCs yielded the best performance for D1. With these optimal choices of model structure and training domain before and after transfer learning, we show the benefits of transfer learning in two experiments.

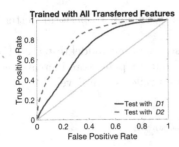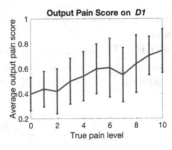

Fig. 11. ROC Curves for classification on NEW test domains *D1* and *D2* using our transfer learning model (left) and plot of average model output pain score (with error bars indicating standard deviation) over true pain level (right).

4.1 Test on New Subjects with only iMotions AU Codings

In this section we report on the results from testing our transfer learning method on a new separate dataset (new participants), which contained only automated features. We trained two models, with and without transfer learning, using all the data in Sect. 3 labeled by both iMotions and humans, and tested the model on this new dataset only labeled by iMotions *D1, D2*. (We use italicized domain names to indicate that this is independent test data *D1, D2*.) Our model with transfer learning (AUC = 0.72±0.002) performed better than the model without it (AUC = 0.67 ± 0.002) on *D1* with a p-value= $1.33e - 45$ in a one-tailed two-sample t-test.

Similar to Fig. 9, in Fig. 11 we plot ROC curves for classification on the NEW test dataset (LEFT) and output pain scores at 0–10 pain levels (RIGHT) using our transfer learning model.

In Fig. 12, we show a scatter plot of neural network output pain scores using transferred automated features versus those using original automated features, as well as pain score distributions, separately for training (All Data from Sect. 3) and test (*D1* from NEW test data in the current section), pain and no-pain. We can see for original automated features scores, no-pain samples from *D1* are distributed very differently from no-pain in All data domain used for training and fall mostly in the range of the pain class. Results using transfer learning do not appear to have this problem.

4.2 Test with Masked Pain and Faked Pain

As another test of the effect of our transfer learning model, we looked at results of classifying whether participants are in pain or not from videos where children were asked to fake pain when they were not really in pain as well as when they were asked to suppress visual expressions of pain when they were in pain.

Although facial expressions convey rich and objective information about pain, they can be deceptive because people can inhibit or exaggerate their pain displays when under observation [21]. It has been shown that human observers discriminate real expressions of pain from faked expressions only marginally better

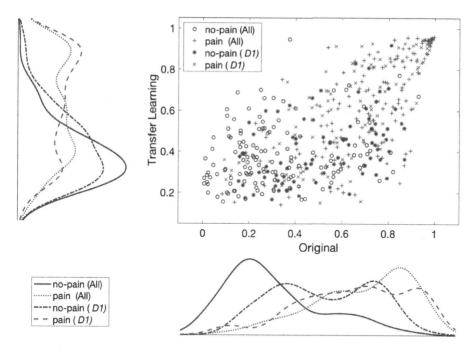

Fig. 12. Scatter plot and distributions of pain scores (transfer learning vs original) using original iMotions features (on the x-axis) and transferred iMotions features (on the y-axis).

than chance [21,22]. Children can also be very good at suppressing pain, but not fully successful in faking expressions of pain [23]. In this section we discuss performance of masked and faked pain in machine learning models trained to distinguish genuine pain and no-pain.

In addition to the data described in Sect. 2.2, we recorded videos of "masked pain" in V1 and V2 by asking participants to suppress pain during the 10-second manual pressure, and videos of "faked pain" during V3 by asking participants to fake the worst pain ever during manual pressure. As in Sect. 2.2, we asked participants to rate their true pain level during manual pressure with a number from 0 to 10. We then labeled masked-pain videos with pain ratings of 4–10 as masked-pain and faked-pain videos with pain ratings of 0–3 as faked-pain, and discarded other samples. This ensured that in masked-pain videos participants actually experienced pain and in faked-pain videos participants in fact felt no pain. One hundred and seventeen masked-pain samples and 116 faked-pain samples were collected. The distribution of the four classes within the three visits is shown in Fig. 13.

Using the best models before and after transfer learning trained to distinguish between genuine pain and no-pain described above, the masked and faked pain samples were processed to obtain pain labels. The results are shown in Fig. 14. We can see that without transfer learning (LEFT), most masked-pain

	Visit 1 and Visit 2 (in hospital)	Visit 3 (in outpatient lab)
Genuine Expression (All Data)	Real Pain	No Pain
Non-genuine Expression	Masked Pain	Faked Pain

Fig. 13. Distribution of four classes in three visits. The area of category is not proportional to the number of samples.

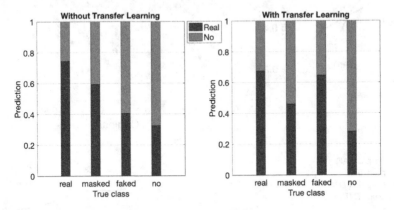

Fig. 14. Bar graph showing classification of real-pain, masked-pain, faked-pain and no-pain. The area of bars shows the distribution of predicting pain and no-pain.

data were classified as real-pain and most faked-pain as no-pain. This appeared to be the case because the AU features coded automatically were sensitive to environmental factors, and during training the machine learned to discriminate between genuine pain and no-pain by recognizing environmental differences between them. At test time, since masked-pain is in the same environmental domain as real-pain and faked-pain is in the similar environment as no-pain, they are assigned to the corresponding classes. In contrast, with transfer learning (Fig. 14 (RIGHT)), masked-pain was mostly classified as no-pain and faked-pain as real-pain. This might be because automated features were transferred to ignore the difference between the two classes caused by environmental change, and the machine can only use differences in facial actions to complete the classification task. Humans' attempts to mask pain are to mimic no-pain faces and, similarly, humans' attempts to fake pain are to mimic pain faces. The machine in this way classifies pain and no-pain according to expressed facial actions.

5   Conclusion

In the present work we recognized differences in classifier model performance (pain vs no-pain) across domains that reflect environmental differences as well as differences reflecting how the data were encoded (automatically v. manually). We demonstrate that manually coded facial features are more robust than automatically coded facial features to environmental changes which allow us to obtain the best performance on our target data domain. We then introduced a transfer learning model to map automated features first to manual pain-related features (or principal components of them), and then used the transferred features for classification (6-2 or 5-4 in Fig. 10). This allowed us to leverage data from another domain to improve classifier performance on the clinically relevant task of automatically distinguishing pain levels in the hospital. Further, we were able to demonstrate improved classifier performance on a separate, new data set.

6   Future Work

Planned future work:

1. Classification of real-pain, masked-pain, faked-pain, and no-pain using machine learning, and comparison to human judgments.
2. Classification of genuine expression and non-genuine expression using machine learning, and comparison to human judgments.
3. Using transfer learning to improve fusion analysis of video features and peripheral physiological features in [24].
4. Multidimensional pain assessment such as pain catastrophizing and anxiety based on facial activities.

Acknowledgments. This work was supported by National Institutes of Health National Institute of Nursing Research grant R01 NR013500, NSF IIS 1528214, and by IBM Research AI through the AI Horizons Network. Many thanks to Ryley Unrau for manual FACS coding and Karan Sikka for sharing his code and ideas used in [5].

References

1. West, J., Ventura, D., Warnick, S.: Spring research presentation: a theoretical foundation for inductive transfer. Brigh. Young Univ. Coll.E Phys. Math. Sci. **1** (2007)
2. Quinn, B.L., Seibold, E., Hayman, L.: Pain assessment in children with special needs: a review of the literature. Except. Child. **82**(1), 44–57 (2015)
3. Zamzmi, G., Pai, C.-Y., Goldgof, D., Kasturi, R., Sun, Y., Ashmeade, T.: Machine-based multimodal pain assessment tool for infants: a review. *preprint* arXiv:1607.00331 (2016)
4. Von Baeyer, C.L.: Children's self-report of pain intensity: what we know, where we are headed. Pain Res. Manag. **14**(1), 39–45 (2009)
5. Sikka, K., et al.: Automated assessment of children's postoperative pain using computer vision. Pediatrics **136**(1), e124–e131 (2015)
6. Aung, M., et al.: The automatic detection of chronic pain-related expression: requirements, challenges and the multimodal emopain dataset. IEEE Trans. Affect. Comput. **7**(4), 435–451 (2016)
7. Sekhon, K.K., Fashler, S.R., Versloot, J., Lee, S., Craig, K.D.: Children's behavioral pain cues: implicit automaticity and control dimensions in observational measures. Pain Res. Manag. (2017)

8. Grunau, R.V.E., Craig, K.D.: Pain expression in neonates: facial action and cry. Pain **28**(3), 395–410 (1987)

9. Hadjistavropoulos, T., et al.: Pain assessment in elderly adults with dementia. Lancet Neurol. **13**(12), 1216–1227 (2014)

10. Ekman, P., Friesen, W.V.: Measuring facial movement. Environ. Psychol. Nonverbal Behav. **1**(1), 56–75 (1976)

11. Martinez, B., Valstar, M.F., Jiang, B., Pantic, M.: Automatic analysis of facial actions: a survey. IEEE Trans. Affect. Comput. (2017)

12. Ashraf, A.B., et al.: The painful face-pain expression recognition using active appearance models. Image Vis. Comput. **27**(12), 1788–1796 (2009)

13. Monwar, M.M., Rezaei, S.: Pain recognition using artificial neural network. In: 2006 IEEE International Symposium on Signal Processing and Information Technology, pp. 28–33. IEEE (2006)

14. Pan, S.J., Yang, Q.: A survey on transfer learning. IEEE Trans. Knowl. Data Eng. **22**(10), 1345–1359 (2010)

15. Xu, X., et al.: Automated pain detection in facial videos of children using human-assisted transfer learning. In: Joint Workshop on Artificial Intelligence in Health, pp. 10–21. CEUR-WS (2018)

16. Hawley, K., et al.: Youth and parent appraisals of participation in a study of spontaneous and induced pediatric clinical pain. Ethics Behav., 1–15 (2018)

17. Hoffman, D.L., Sadosky, A., Dukes, E.M., Alvir, J.: How do changes in pain severity levels correspond to changes in health status and function in patients with painful diabetic peripheral neuropathy. Pain **149**(2), 194–201 (2010)

18. Littlewort, G., et al.:. The computer expression recognition toolbox (CERT). In: 2011 IEEE International Conference on Automatic Face & Gesture Recognition and Workshops (FG 2011), pp. 298–305. IEEE (2011)

19. Prkachin, K.M.: The consistency of facial expressions of pain: a comparison across modalities. Pain **51**(3), 297–306 (1992)

20. Prkachin, K.M.: Assessing pain by facial expression: facial expression as nexus. Pain Res. Manag. **14**(1), 53–58 (2009)

21. Hill, M.L., Craig, K.D.: Detecting deception in facial expressions of pain: accuracy and training. Clin. J. Pain **20**(6), 415–422 (2004)

22. Bartlett, M.S., Littlewort, G.C., Frank, M.G., Lee, K.: Automatic decoding of facial movements reveals deceptive pain expressions. Curr. Biol. **24**(7), 738–743 (2014)

23. Larochette, A.-C., Chambers, C.T., Craig, K.D.: Genuine, suppressed and faked facial expressions of pain in children. Pain **126**(1–3), 64–71 (2006)

24. Xu, X., et al.: Towards automated pain detection in children using facial and electrodermal activity. In: Joint Workshop on AI in Health, pp. 208–211. CEUR-WS (2018)

Towards Automated Pain Detection in Children Using Facial and Electrodermal Activity

Xiaojing Xu[1]([✉])(iD), Büşra Tuğçe Susam[2](iD), Hooman Nezamfar[3](iD),
Damaris Diaz[4](iD), Kenneth D. Craig[5](iD), Matthew S. Goodwin[6](iD),
Murat Akcakaya[2](iD), Jeannie S. Huang[4](iD), and Virginia R. de Sa[7](iD)

[1] Department of Electrical and Computer Engineering, UC San Diego,
La Jolla, CA, USA
xix068@ucsd.edu
[2] Department of Electrical and Computer Engineering, University of Pittsburgh,
Pittsburgh, PA, USA
[3] Department of Electrical and Computer Engineering, Northeastern University,
Boston, MA, USA
[4] Rady Childrens Hospital and Department of Pediatrics, UC San Diego, La Jolla,
CA, USA
[5] Department of Psychology, University of British Columbia, Vancouver, BC, Canada
[6] Department of Health Sciences, Northeastern University, Boston, MA, USA
[7] Department of Cognitive Science, UC San Diego, La Jolla, CA, USA

Abstract. Accurately determining pain levels in children is difficult, even for trained professionals and parents. Facial activity and electrodermal activity (EDA) provide rich information about pain, and both have been used in automated pain detection. In this paper, we discuss preliminary steps towards fusing models trained on video and EDA features respectively. We compare fusion models using original video features and those using transferred video features which are less sensitive to environmental changes. We demonstrate the benefit of the fusion and the transferred video features with a special test case involving domain adaptation and improved performance relative to using EDA and video features alone.

Keywords: Automated pain detection · Domain adaptation · EDA · Facial action units · GSR · FACS

1 Introduction

Accurate pain assessment in children is necessary for safe and efficacious pain management. Under-estimation can lead to patient suffering and inadequate care, while over-estimation can lead to overdosing of pain medication, which may predispose other issues, including opioid addiction [1]. The most widely used method of assessing clinical pain is patient self-report [2]. However, this

© Springer Nature Switzerland AG 2019
F. Koch et al. (Eds.): AIH 2018, LNAI 11326, pp. 181–189, 2019.
https://doi.org/10.1007/978-3-030-12738-1_13

method is subjective and vulnerable to social and self-presentation biases and requires substantial cognitive, linguistic, and social competencies [2]. Objective pain estimation is required for appropriate pain management in the clinical setting.

In previous work, features extracted from facial action units (AUs) and electrodermal activity (EDA) signals have both been used to automatically detect pain events using machine learning methods, and transfer learning techniques have been applied to environmentally sensitive features like facial AU features for domain adaptation tasks [3–6]. In this work, we design a machine learning model that utilizes both facial and EDA features to recognize pain. We trained a unimodal model using either facial or EDA features to give pain scores, and then trained another fusion model using scores from the two models. We also applied transfer learning [6] to transfer video features into a domain-robust space, and analyzed the fusion model using transferred video features combined with EDA features. We observed performance improvement in a domain adaptation task with the fusion method.

An earlier and shorter version of this work was presented at the Joint Workshop on Artificial Intelligence in Health and appears in the proceedings [7].

2 Methods

2.1 Participants

Forty-two pediatric research participants (30 males, 12 females) aged 13 [10, 15] (median [25%, 75%]) years and primarily Hispanic (79%) who had undergone medically necessary laparoscopic appendectomy were recruited for a study examining automated assessment of children's post-operative pain using video and wearable biosensors. Children and their parents provided assent and parental consent prior to study evaluations [8].

2.2 Experimental Design and Data Collection

Data were collected over 3 visits (V): (V1) within 24 h after appendectomy in hospital; (V2) in hospital one calendar day after V1; and (V3) a follow-up visit in an outpatient lab up to 42 days postoperatively. At each visit, videos (60 fps at 853×480 pixel resolution) of the participant's face and EDA responses (using Affectiva Q sensor) were recorded while manual pressure was exerted at the surgical site for 10 s (equivalent of a clinical examination). During hospital visit (V1, V2) data collections, participants were lying in a hospital bed with the head of the bed raised. In V3, they were seated in a reclined chair. Participants rated their pain level using a 0–10 Numerical Rating Scale, where 0 = no pain and 10 = worst pain ever. Following convention for recognizing clinically significant pain [9], videos and EDA with ratings of 0–3 were labeled as no-pain, and videos and EDA with ratings of 4–10 were labeled as pain. We obtained 22 pain samples from V1, 8 pain and 8 no-pain samples from V2, and 22 no-pain samples from V3. The data distribution is illustrated in Fig. 1.

V1 (in hospital)	V2 (in hospital)		V3 (in outpatient lab)
22 pain videos	8 pain videos	8 no-pain videos	22 no-pain videos

Fig. 1. Data distribution over three visits and two environmental domains.

2.3 Feature Extraction and Processing

Video Features: Each 10-second video during pressure was processed with iMotions software which automatically estimates the log probabilities of 20 AUs (AU 1, 2, 4, 5, 6, 7, 9, 10, 12, 14, 15, 17, 18, 20, 23, 24, 25, 26, 28, 43) and 3 head pose indicators (yaw, pitch and roll) from each frame. We then applied 11 statistics (mean, max, min, standard deviation, 95th, 85th, 75th, 50th, 25th percentiles, half-rectified mean, and max-min) to each AU over all frames to obtain 11×23 features.

EDA Features: EDA signals were trimmed to 30 s (10 s before, during, and after pressure was exerted at the surgical site respectively), smoothed by a 0.35 Hz FIR low pass filter, down-sampled to 1 Hz, and normalized with z-score normalization. We then applied timescale decomposition (TSD) with a standard deviation metric, and computed the mean, SD, and entropy of each row of each TSD to obtain a feature vector of length 90 [5].

Transferred Video Features: In [6], we collected videos from 73 children over 3 visits and had a human expert code videos along with iMotions, and then used the data to learn a mapping from iMotions features to human features. The mapped iMotions features gained the domain robustness property of human features. In this paper, we processed video features through the learned model to get domain-robust transferred video features.

2.4 Machine Learning Models

Support Vector Machine (SVM): A linear SVM was used to obtain a pain score as well as a pain prediction for each sample using transferred video features or video/EDA features after PCA. The number of principal components was chosen using cross-validation and the box constraint is set to 1. Inputs were normalized with z-score normalization over the full dataset as in [5].

Linear Discriminant Analysis (LDA): LDA was used to differentiate between pain and no-pain using output pain scores from the SVMs. Inputs were either one single pain score from one SVM, or a fusion of pain scores from both SVMs.

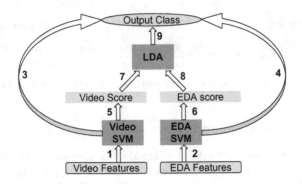

Fig. 2. Graph of model hierarchy.

Neural Networks: Neural Networks with one hidden layer were used to map iMotions features to the human feature subspace. The number of hidden neurons was 506, twice the number of input features. Batch normalization and sigmoid activation were used for the hidden layer. The output layer used the linear activation function and mean squared error as the loss function.

2.5   Evaluation Metrics

Our primary objective was to classify pain in the hospital setting. For this reason we used V2 (in hospital pain vs no-pain) as test data.

We use classification accuracy to quantify performance of the model. Classification accuracy reflects the percentage of correctly classified trials for a given learned threshold. We also report sensitivity and specificity along with overall accuracy to indicate whether the model is better at detecting pain or no-pain. In order to measure how well the classifiers are able to separate pain and no-pain classes, we also use AUC (Area under the ROC curve) which is insensitive to the classification threshold. It measures the area under the curve that plots hit rate vs false alarm rate as the threshold is moved over all possible values. An AUC of 1 reflects perfect separation and an AUC of 0.5 reflects no separation.

3   Results and Discussion

3.1   Performance Using Video/EDA Features

We first used V1 pain and V3 no-pain samples to train an SVM for classification, following $1 \Rightarrow 3$ and $2 \Rightarrow 4$ in Fig. 2. Table 1 shows that SVM performance on V2 was good for EDA (accuracy $= 0.75$), but suboptimal for video features (accuracy $= 0.5$). The AUC for video was 0.66, which is higher than random, implying that the model learned a function output, or score, that correlated with pain self-report scores, but failed to find an appropriate classification threshold.

Table 1. Performance for classification on V2.

	Video	EDA	Video+EDA	Video-V2	EDA-V2	Video+EDA-V2
Acc	0.5	0.75	0.56	0.69	0.71	**0.84**
Sensitivity	0.75	0.75	0.87	0.62	0.62	**0.84**
Specificity	0.25	0.75	0.25	0.75	0.80	**0.83**
AUC	0.66	0.80	**0.88**	–	–	–

In a separate experiment, we trained an SVM with V1 pain and V3 no-pain data, and tested the model with V1 pain and V3 no-pain data. The classification accuracy was 0.8 for video features, much higher than performance tested on V2 pain and no-pain. We hypothesize that the observed difference was due to iMotions feature sensitivity to environmental differences between V1/2 (in hospital) and V3 (in outpatient lab). When training with V1 pain and V3 no-pain, the classifier learns the difference between the hospital environment of V1 and outpatient lab environment of V3, but in testing, such difference no longer exists between the two classes in V2, so the classification fails. This problem has previously been discussed in [6].

One solution to this problem was to use V2 to train the model. However, with only 16 data points in V2, results had large variance. Likewise, training with V1/3 and V2 data together did not improve V2 performance. Consequently, we needed to solve the domain adaptation problem which learns a model from a source domain (V1/3) and applies it on a different target domain (V2).

3.2 Fusion of Video and EDA

We hoped to improve performance on V2 by combining video and EDA features. Our first simple attempt at fusion was to fit an LDA model to distinguish between pain v. no-pain using the output pain scores from each of the SVM models trained with video and EDA features respectively ($1, 2 \Rightarrow 5, 6 \Rightarrow 7, 8 \Rightarrow 9$ in Fig. 2). However, this method performed poorly (accuracy = 0.56 under "Video+EDA" column in Table 1) compared to using EDA features alone. Nevertheless, the AUC of this model was 0.88, higher than the AUCs of both video and EDA alone, showing potential for the fusion model. In the next section, we demonstrate a method that better adjusts the decision boundary of the fusion model to get a higher classification accuracy.

3.3 Training with V2 Scores

Through fusion, our LDA classifier involved only two inputs: video and EDA SVM pain scores. In such structure, the SVM models can be regarded as encoders or feature extractors. Since the dimensionality of features was greatly reduced by the SVM, it became feasible to train a classifier using only V2 samples. Relative to Fig. 2, we thus trained 1,2 with V1/3, and 7,8 with V2 data using

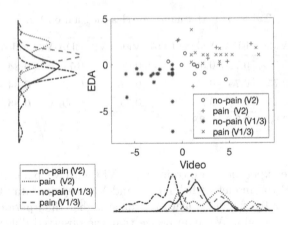

Fig. 3. Scatter plot and distributions for EDA and video pain scores output by SVM models, grouped by data domain and class.

cross-validation. The accuracy using video scores alone as LDA inputs was 0.69, much higher than 0.5, showing the benefit of training on target domain V2, even if the features, V2 scores, were obtained from a model trained on V1/3. Finally, with a fusion of SVM output scores for video and EDA, we achieved a comparative best accuracy of 0.84.

It should be noted that we performed leave-one-sample-out cross-validation on V2 samples in order to train and test both on V2, so AUCs were not measurable.

To better understand our problem, we plot EDA SVM output scores versus video SVM output scores in Fig. 3 for data points from two domains (V1/3 and V2) and two classes (pain and no-pain). We also plot score distributions for EDA and video along corresponding axes. These scores were obtained by SVM trained on V1/3 domain. We can see that for both EDA and video features, pain and no-pain classes are well-separated on V1/3. However, for video features, no-pain on V2 drifts towards pain and will clearly be classified as pain if the threshold differentiating pain and no-pain on V1/3 is used. This supports our hypothesis that the video SVM model learned to classify "in hospital" versus "in outpatient lab" instead of pain versus no-pain during training with V1/3 data, so that V2 no-pain test data are classified by their "in hospital" properties.

As we can see from the video score plot, if we adjust the classification threshold based on V2 pain scores, we obtain improved performance (Table 1 "Video-V2"). The scatter plot on the other hand explains why we should benefit from the fusion of EDA and video since the optimal decision boundary that partitions the two distributions is oblique, which means the decision model should weight both dimensions.

Table 2. Performance for classification on V2 (transfer learning).

	TF	EDA	TF+EDA	TF-V2	EDA-V2	TF+EDA-V2
Acc	0.81	0.75	0.75	0.76	0.71	0.80
Sensitivity	0.87	0.75	0.87	0.77	0.62	0.81
Specificity	0.75	0.75	0.62	0.75	0.80	0.78
AUC	0.81	0.80	**0.88**	–	–	–
				TF-V123	EDA-V123	TF+EDA-V123
Acc				0.81	0.73	**0.86**
Sensitivity				0.87	0.86	**0.87**
Specificity				0.75	0.59	**0.84**
AUC				–	–	–

3.4   Transfer Learning for Video Features

We introduced a transfer learning method in [6] to map automated video features to a subspace of manual pain-related video features which loses sensitivity to domain changes while keeping necessary information to recognize pain. We applied the same method to the video features in our experiment and show the results in Table 2 under "TF" (TF = transferred video feature). Accuracy using video features on target domain V2 improved from 0.5 to 0.81 with transfer learning when the SVM model was trained on source domain V1/3. Unlike original video features, in the plot of score distribution for transferred video features in Fig. 4, no-pain (V2) is distributed similarly to no-pain (V1/3), showing the domain robustness of video features after transfer learning.

3.5   Fusion of Transferred Video and EDA

We then combined transferred video and EDA as we did in Sect. 3.2, following $1, 2 \Rightarrow 5, 6 \Rightarrow 7, 8 \Rightarrow 9$ in Fig. 2 where we replaced video by transferred video. The accuracy is 0.75 under "TF+EDA", no higher than using TF or EDA alone, but the AUC (0.88) beats both (Table 2).

In Sect. 3.3 we retrained the LDA model on V2 to improve accuracy. Since transferred video features and EDA features are both invariant to domain changes, we did not have to exclude V3 during training. Instead we could leave one sample in V2 test data out at each iteration, use the remaining V2 samples and all V1/3 samples together for training, and produce a decision for each V2 sample. The accuracy of this fusion model was 0.86. To compare with Table 1, we also report accuracies training LDA with V2 in Table 2. The accuracies are slightly lower than training with V1-3 together, possibly due to decreased training sample size.

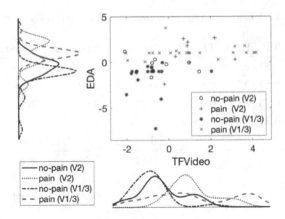

Fig. 4. Scatter plot and distributions for EDA and transferred video pain scores output by SVM models, grouped by data domain and class.

4   Conclusion

We present preliminary results using a fusion approach to automatically detect pain in children. While the results demonstrate improvement with our domain adaptation fusion methodology over approaches using video, transferred video, or EDA features alone, we believe these results can be further improved by tailoring the two modalities to be more sensitive to their relative benefits and limitations.

Acknowledgments. This work was supported by National Institutes of Health National Institute of Nursing Research grant R01 NR013500 and NSF IIS 1528214.

References

1. Quinn, B.L., Seibold, E., Hayman, L.: Pain assessment in children with special needs: a review of the literature. Except. Child. **82**(1), 44–57 (2015)
2. Zamzmi, G., Pai, C.-Y., Goldgof, D., Kasturi, R., Sun, Y., Ashmeade, T.: Machine-based multimodal pain assessment tool for infants: a review. arXiv:1607.00331 (2016)
3. Sikka, K., et al.: Automated assessment of children's postoperative pain using computer vision. Pediatrics **136**(1), e124–e131 (2015)
4. Gruss, S., et al.: Pain intensity recognition rates via biopotential feature patterns with support vector machines. PloS One **10**(10), e0140330 (2015)
5. Susam, B.T., et al.: Automated pain assessment using electrodermal activity data and machine learning. In: 2018 40th Annual International Conference of the IEEE Engineering in Medicine and Biology Society (EMBC). IEEE (2018, in Press)
6. Xu, X., et al.: Automated pain detection in facial videos of children using human-assisted transfer learning. In: Joint Workshop on Artificial Intelligence in Health, pp. 10–21. CEUR-WS (2018)

7. Xu, X., et al.: Towards automated pain detection in children using facial and electrodermal activity. In: Joint Workshop on AI in Health, pp. 208–211. CEUR-WS (2018)
8. Hawley, K., et al.: Youth and parent appraisals of participation in a study of spontaneous and induced pediatric clinical pain. Ethics Behav. 1–15 (2018)
9. Hoffman, D., Sadosky, A., Dukes, E., Alvir, J.: How do changes in pain severity levels correspond to changes in health status and function in patients with painful diabetic peripheral neuropathy. Pain **149**(2), 194–201 (2010)

Interpretation of Best Medical Coding Practices by Case-Based Reasoning—A User Assistance Prototype for Data Collection for Cancer Registries

Michael Schnell[1,2(✉)], Sophie Couffignal[1], Jean Lieber[2], Stéphanie Saleh[1], and Nicolas Jay[2,3]

[1] Department of Population Health, Luxembourg Institute of Health,
1A-B, rue Thomas Edison, 1445 Strassen, Luxembourg
{michael.schnell,sophie.couffignal,stephanie.saleh}@lih.lu
[2] UL, CNRS, Inria, Loria, 54000 Nancy, France
jean.lieber@loria.fr
[3] Service d'évaluation et d'information médicales,
Centre Hospitalier Régional Universitaire de Nancy, Nancy, France
n.jay@chru-nancy.fr

Abstract. In the fight against cancer, cancer registries are an important tool. At the heart of these registries is the data collection and coding process. This process is ruled by complex international standards and numerous best practices, which can easily overwhelm (coding) operators. In this paper, a system assisting operators in the interpretation of best medical coding practices and a short evaluation are presented. By leveraging the arguments used by the coding experts to determine the best coding option, the proposed system answers coding questions from operators and provides a partial explanation for the proposed solution.

Keywords: Interpretation of best practices ·
Interpretive case-based reasoning · Coding standards ·
Cancer registries · User assistance · Decision support

1 Introduction

There are numerous cancer registries around the world collecting data about cancers diagnosed and/or treated in a given area. This data is used to monitor cancer (incidence rates, survival rates, etc.) and to evaluate cancer care (diagnosis, treatment, etc.). To produce comparable data, common definitions (e.g. terminologies like the International Classification of Diseases (ICD)) and coding practices [5] have to be followed. However, the broadness and complexity of these standards make the work of the medical staff in charge of coding (operators) more difficult.

© Springer Nature Switzerland AG 2019
F. Koch et al. (Eds.): AIH 2018, LNAI 11326, pp. 190–198, 2019.
https://doi.org/10.1007/978-3-030-12738-1_14

The aim of this research is to address this complexity, by assisting both operators and coding experts in the interpretation of coding best practices.

As an illustrating example, let us consider the case denoted by `exmpl` of a particular woman. In 2016, multiple pulmonary opacities were discovered within her right lung. A CT scan indicated no mediastinal adenopathy.[1] A histological analysis of a sample identified the morphology[2] of the cancer as adenocarcinoma. The TTF1 marker test was positive. After further testing, another tumor is found in the ovaries. An operator might wonder which topography[3] should be coded (lung or ovaries?) and can request help. For the Luxembourg National Cancer Registry (NCR), operators ask their questions using an online ticketing system. With free text description provided by operators, coding experts provide a solution, i.e. an answer with their reasoning in the form of a motivated argument.

Section 2 describes an approach to assist the data collection process for cancer registries and how case-based reasoning (CBR [1]) is applied. In Sect. 3, a prototype and preliminary results are discussed. Section 4 presents a conclusion and points out what further efforts need to be undertaken in the future.

2 Case-Based Interpretation of Best Practices

This article summarizes the work presented in [9] and adds a description of the developed prototype and some preliminary results.

2.1 Preliminaries

RDFS[4] is a knowledge representation language of the semantic web. SPARQL (See footnote 4) is a query language for RDFS web.

A case $(\mathtt{srce}, \mathtt{sol}(\mathtt{srce}))$ is composed of two parts: (1) `srce` is a problem given by a question (i.e. a subject) and a patient record, and (2) `sol(srce)` a solution for the problem `srce`.

The question indicates the subject (incidence date, topography, tumor nature, etc.). In the example, the question is about the topography.

The patient record represents the data from the hospital patient record (patient features, tumors, exams, treatments, etc.) needed to answer the question. The relevant data depends on the subject and is defined by coding experts. The patient record is represented by an RDFS graph [3] (see Fig. 1). Body parts and cancer morphologies use classes from the SNOMED Clinical Terms[5] ontology.

---

[1] An adenopathy is an enlargement of lymph nodes, likely due to cancer.

[2] The morphology describes the type and behavior of the cells that compose the tumor.

[3] The topography is the location where the tumor originated.

[4] https://www.w3.org/TR/rdf-schema/   and   https://www.w3.org/TR/sparql11-query.

[5] https://bioportal.bioontology.org/ontologies/SNOMEDCT.

The solution contains the answer to the question and the most important arguments in favor of (**pros**) and against (**cons**) this answer. In the example, the answer is to consider the topography to be the ovaries. The presence of multiple pulmonary opacities is an argument in favor, as they are indicative of a lung metastasis and thus the tumor is unlikely to have originated in the lungs.

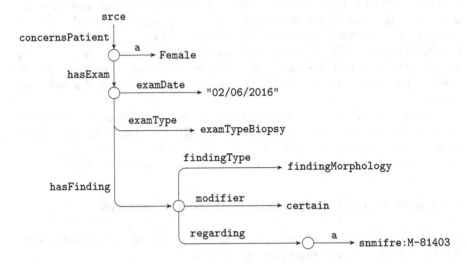

Fig. 1. Short patient record in RDFS. This graph represents a woman with a single biopsy (exam), identifying the tumor as adenocarcinoma (which is coded as M-8140/3). The circles represent blank nodes.

The arguments have two uses. They help explain the answer to operators and serve as a reminder for coding experts. They are also used in the proposed approach during the retrieval step. Three types of arguments will be considered: strong pros, weak pros and weak cons. The difference between a strong and a weak argument comes from their reliability for a given conclusion. A strong argument is considered to be a sufficient justification for an answer, unlike a weak argument which is more of an indication or clue. It can be noted that there are no strong cons in the source cases. Indeed, such an argument would be an absolute argument against the given answer. Formally, an argument is a function that associates a Boolean to a case and is stored as a SPARQL ASK query. The following shows an argument `arg`, followed by an explanation:

$$\text{arg(case)} = \begin{array}{|l} \texttt{ASK \{} \\ \quad \texttt{case concernsPatient ?patient .} \\ \quad \texttt{?patient hasExam ?exam\_morpho .} \\ \quad \texttt{?exam\_morpho hasFinding ?finding .} \\ \quad \texttt{?finding findingType findingTypeFindMorphology .} \\ \quad \texttt{?finding modifier certain .} \\ \quad \texttt{?finding regarding [a snmifr:M-81403] .} \\ \quad \texttt{?patient hasExam ?exam\_ttf .} \\ \quad \texttt{?exam\_ttf hasFinding ?finding .} \\ \quad \texttt{?finding findingType findingTypeFindTTF1Marker .} \\ \quad \texttt{?finding present yes .} \\ \texttt{\}} \end{array}$$

arg says that a TTF1 positive adenocarcinoma is in favor of a primitive lung cancer. The argument checks that the morphology of the tumor is of type adenocarcinoma and that the tumor is positive for the TTF1 marker. This argument applies for the example described in the introduction, i.e. arg(exmpl) = TRUE.

2.2   Global Architecture

The proposed approach uses a 4-R cycle (retrieve, reuse, revise, retain) adapted from [1] and four knowledge containers [8] (case base, domain knowledge, retrieval knowledge, adaptation knowledge), as shown in Fig. 2.

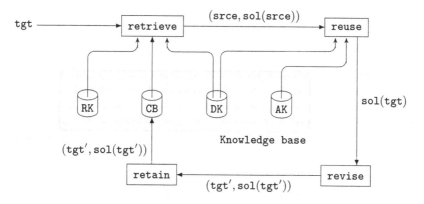

Fig. 2. Adapted 4-R cycle and knowledge containers for the proposed approach.

2.3   Retrieve

The proposed approach relies on arguments to find similar cases. Indeed, similar answers should be based on similar reasoning and thus the same arguments should apply. Our method checks the applicability of arguments from source cases on the target problem `tgt` and uses this to determine the preferred source case to solve `tgt`. This preference relation is denoted by the preorder $\preceq_{\texttt{tgt}}$. The comparison between two source cases `i` and `j` relies on three criteria, C_s for strong arguments, C_w for weak arguments and $C_{\texttt{dist}}$ for patient records.

An argument `arg` is applicable for a case `c` if the preconditions of the argument are met in the patient record of `c`. For the argument `arg` described in the preliminaries, `arg` applies for a case if the patient record contains at least two exams, one identifying the morphology as adenocarcinoma and another exam reporting a TTF1 positive tumor. Formally an argument `arg` is applicable for a case c if `arg(c) = TRUE`.

For the criterion C_s, the source case with more applicable strong arguments is preferred. Formally, C_s is met if $\Delta_{i,j}^s > 0$, where $\Delta_{i,j}^s$ is defined as

$$\Delta_{i,j}^s = \mathcal{N}^{\texttt{sp}}(\texttt{srce}_i, \texttt{tgt}) - \mathcal{N}^{\texttt{sp}}(\texttt{srce}_j, \texttt{tgt})$$

and $\mathcal{N}^{\texttt{args}}(\texttt{srce}, \texttt{tgt})$ denotes the number of arguments of type `args` of a the source case `srce` which are applicable for a case `tgt`, i.e.

$$\mathcal{N}^{\texttt{args}}(\texttt{srce}, \texttt{tgt}) = |\{\texttt{a} \in \texttt{args}(\texttt{srce}) \mid \texttt{a}(\texttt{tgt}) = \texttt{TRUE}\}|$$

and $\texttt{args} \in \{\texttt{sp}, \texttt{wp}, \texttt{wc}\}$ is an argument type. `sp(srce)` is the set of strong pros, `wp(srce)` the set of weak pros and `wc(srce)` the set of weak cons of `srce`.

For the criterion for weak arguments C_w, a combination of pros and cons is used. Intuitively, if more weak pros and less weak cons are applicable, the source case is preferred. Formally, C_w is met if $\Delta_{i,j}^w > 0$, where $\Delta_{i,j}^w$ is defined as

$$\begin{aligned}\Delta_{i,j}^w = \ & \lambda_p * (\mathcal{N}^{\texttt{wp}}(\texttt{srce}_i, \texttt{tgt}) - \mathcal{N}^{\texttt{wp}}(\texttt{srce}_j, \texttt{tgt})) \\ & - \lambda_c * (\mathcal{N}^{\texttt{wc}}(\texttt{srce}_i, \texttt{tgt}) - \mathcal{N}^{\texttt{wc}}(\texttt{srce}_j, \texttt{tgt}))\end{aligned}$$

where λ_p and λ_c are two nonnegative coefficients that are currently fixed to $\lambda_p = 3$ and $\lambda_c = 2$. When more data are available, these parameters values will be reevaluated.

For the criterion $C_{\texttt{dist}}$, a graph edit distance between patient record RDFS graphs is used [4]. Formally, $C_{\texttt{dist}}$ is met if $\Delta_{i,j}^{\texttt{dist}} \geq 0$, where $\Delta_{i,j}^{\texttt{dist}}$ is defined as

$$\Delta_{i,j}^{\texttt{dist}} = \texttt{dist}(\texttt{srce}_j, \texttt{tgt}) - \texttt{dist}(\texttt{srce}_i, \texttt{tgt})$$

The three criteria are considered lexicographically, first C_s, then C_w and finally $C_{\texttt{dist}}$ (see [9]). $\texttt{srce}_i$ is preferred over $\texttt{srce}_2$, i.e $\texttt{srce}_i \preceq_{\texttt{tgt}} \texttt{srce}_j$, ifq

$$\Delta_{i,j}^s > 0 \text{ or } (\Delta_{i,j}^s = 0 \text{ and } (\Delta_{i,j}^w > 0 \text{ or } (\Delta_{i,j}^w = 0 \text{ and } \Delta_{i,j}^{\texttt{dist}} \geq 0)))$$

2.4   Reuse

Once an appropriate source case has been found, the solution associated to the source case is copied: `sol(tgt) := sol(srce)`. The arguments that do not apply to the target problem, if any, are removed.

2.5   Revise and Retain

The newly formed case $(\texttt{tgt}, \texttt{sol}(\texttt{tgt}))$ can be reviewed by a coding expert, to modify the answer, the arguments and/or the patient record. A coding expert may choose to remove unnecessary information from the patient record, removing unwanted specificity. Thus, $(\texttt{tgt}, \texttt{sol}(\texttt{tgt}))$ is substituted by $(\texttt{tgt}', \texttt{sol}(\texttt{tgt}'))$, where $\texttt{tgt}'$ is more general than $\texttt{tgt}$. $(\texttt{tgt}', \texttt{sol}(\texttt{tgt}'))$ is a generalized case that has a larger coverage than $(\texttt{tgt}, \texttt{sol}(\texttt{tgt}))$ [6].

3   Prototype and Preliminary Results

The prototype designed for the NCR serves as a ticketing system, where operators ask coding questions and experts provide answers. It assists operators in structuring questions, making it easier for the NCR and coding experts to find similar questions later. For topography questions, it will also provide a tentative answer. This answer is calculated using the approach described in [9]. All the answers are reviewed by experts. The prototype presents itself as a single page

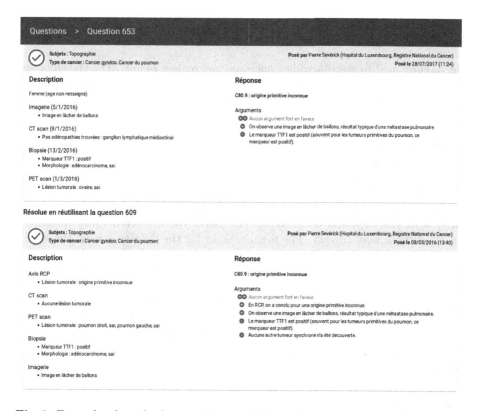

Fig. 3. Example of a solved case. The top displays the new question asked and the provided solution. The bottom displays the source case used to solve the new question.

application built using Angular[6] with a backing REST API built with Go (See footnote 6) and the Gin framework.[7] The data is stored in a triple store Apache Jena and exposed as a SPARQL endpoint using Apache Fuseki.[8] Figures 3 and 4 show screenshots of the prototype.

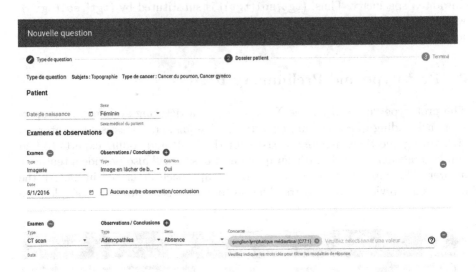

Fig. 4. Form used to describe coding questions and patient records. The French labels for body parts and morphologies are taken from the SNMIFRE (a French translation of SNOMED, http://bioportal.lirmm.fr/ontologies/SNMIFRE).

The prototype was tested internally, to perform a first assessment of its usability and utility. Some old cases concerning the topography were formalized and coded, with some domain knowledge. For the arguments, great care was given during modeling in order to make them more broadly applicable. Then new questions were presented to the system, and the proposed solution compared with the expected ones. While the prototype answered every question, not all of them were correct. The main reasons for the difference were the small amount of cases (15 originally, however the case base will be enriched by routine usage) and the simple reuse method used at this stage. Indeed, as the arguments have been formalized to be more general, some of the provided answers might be slightly incorrect (e.g. answering upper lung lobe instead of lower lung lobe). Despite this, as the prototype displays the reused source case, an operator should be able to make the necessary adaptation to the provided solution. For the questions concerning other subjects, the prototype relies entirely on the coding experts to provide answers.

---

[6] https://angular.io.

[7] https://golang.org, https://github.com/gin-gonic/gin.

[8] https://jena.apache.org/ and https://jena.apache.org/documentation/fuseki2/.

To the best of our knowledge, few other research attempts to use arguments in the context of the retrieval process. The closest method found is a work by McSherry [7]. The proposed approach creates explanations afterwards, using the closest source case to provide the conclusion and the closest source case with the opposite conclusion to compute which attributes favor the conclusion and which attributes do not. Unlike our approach, each argument is linked to a single attribute. Thus they cannot show how the combination of attributes might influence a given outcome.

4 Conclusion

Recently there has been a growing interest for case-based reasoning applications in health sciences [2]. In this paper, an approach to assist operators in the interpretation of best medical coding practices has been proposed. This approach is based on discussions with operators and coding experts on actual coding problems. A dozen tricky problems were discussed in detail, among a hundred simpler problems. The coding questions asked by the operators are compared to previous questions and solved by reusing the pros and cons of previously given solutions. The results discussed are only preliminary and a more thorough evaluation, including the operators and coding experts, is planned.

At the moment the reasoning process is only partial. Arguments are only a part of a more complex reasoning process. The formalization of this process and the eventual integration of the coding standards remains an interesting avenue for future work.

After the prototype has been validated and improved by routine usage, a second version will be designed that is less domain-dependent. The objective is to build a generic system for argumentative case-based reasoning using semantic web standards.

Acknowledgments. The authors wish to thank the anonymous reviewers of the Joint Workshop on Artificial Intelligence in Health for their remarks which have helped in improving the quality of the paper. The first author would also like to thank the Fondation Cancer for their financial support.

References

1. Aamodt, A., Plaza, E.: Case-based reasoning: foundational issues, methodological variations, and system approaches. AI Commun. **7**, 39–59 (1994)
2. Bichindaritz, I., Marling, C., Montani, S.: Case-based reasoning in the health sciences. In: Workshop Proceedings of ICCBR (2015)
3. Brickley, D., Guha, R.V.: RDF Schema 1.1, W3C recommendation, last consultation: March 2017 (2014). https://www.w3.org/TR/rdf-schema/
4. Bunke, H., Messmer, B.T.: Similarity measures for structured representations. In: Wess, S., Althoff, K.-D., Richter, M.M. (eds.) EWCBR 1993. LNCS, vol. 837, pp. 106–118. Springer, Heidelberg (1994). https://doi.org/10.1007/3-540-58330-0_80

5. Tyczynski, J.E., Démaret, E., Parkin, D.M., European Network of Cancer Registries: Standards and Guidelines for Cancer Registration in Europe: the ENCR Recommendations. International Agency for Research on Cancer, Lyon (2003)

6. Maximini, K., Maximini, R., Bergmann, R.: An investigation of generalized cases. In: Ashley, K.D., Bridge, D.G. (eds.) ICCBR 2003. LNCS (LNAI), vol. 2689, pp. 261–275. Springer, Heidelberg (2003). https://doi.org/10.1007/3-540-45006-8_22

7. McSherry, D.: Explaining the pros and cons of conclusions in CBR. In: Funk, P., González Calero, P.A. (eds.) ECCBR 2004. LNCS (LNAI), vol. 3155, pp. 317–330. Springer, Heidelberg (2004). https://doi.org/10.1007/978-3-540-28631-8_24

8. Richter, M.M., Weber, R.O.: Case-Based Reasoning: A Textbook. Springer, Berlin (2013). https://doi.org/10.1007/978-3-642-40167-1

9. Schnell, M., Couffignal, S., Lieber, J., Saleh, S., Jay, N.: Case-based interpretation of best medical coding practices—application to data collection for cancer registries. In: Aha, D.W., Lieber, J. (eds.) ICCBR 2017. LNCS (LNAI), vol. 10339, pp. 345–359. Springer, Cham (2017). https://doi.org/10.1007/978-3-319-61030-6_24

Identification of Serious Illness Conversations in Unstructured Clinical Notes Using Deep Neural Networks

Isabel Chien[1] , Alvin Shi[1], Alex Chan[2], and Charlotta Lindvall[3,4(✉)]

[1] Massachusetts Institute of Technology, Cambridge, USA
{chieni,alvinshi}@mit.edu
[2] Harvard T.H. Chan School of Public Health, Boston, USA
alexchan@mail.harvard.edu
[3] Dana-Farber Cancer Institute, Boston, USA
[4] Brigham and Women's Hospital, Boston, USA
clindvall@mail.harvard.edu

Abstract. Advance care planning, which includes clarifying and documenting goals of care and preferences for future care, is essential for achieving end-of-life care that is consistent with the preferences of dying patients and their families. Physicians document their communication about these preferences as unstructured free text in clinical notes; as a result, routine assessment of this quality indicator is time consuming and costly. In this study, we trained and validated a deep neural network to detect documentation of advanced care planning conversations in clinical notes from electronic health records. We assessed its performance against rigorous manual chart review and rule-based regular expressions. For detecting documentation of patient care preferences at the note level, the algorithm had high performance; F1-score of 92.0 (95% CI, 89.1–95.1), sensitivity of 93.5% (95% CI, 90.0%–98.0%), positive predictive value of 90.5% (95% CI, 86.4%–95.1%) and specificity of 91.0% (95% CI, 86.4%–95.3%) and consistently outperformed regular expression. Deep learning methods offer an efficient and scalable way to improve the visibility of documented serious illness conversations within electronic health record data, helping to better quality of care.

Keywords: Deep learning · End-of-life care · Palliative care · Natural language processing · Clinical notes · Electronic health records

1 Introduction and Related Work

To ensure that patients receive care that is consistent with their goals, clinicians must communicate with seriously ill patients about their treatment preferences. More than 80% of Americans say they would prefer to die at home, if possible. Despite this, 60% of Americans die in acute care hospitals and 20% die in an Intensive Care Unit (ICU) [1]. Advance care planning, which includes clarifying

© Springer Nature Switzerland AG 2019
F. Koch et al. (Eds.): AIH 2018, LNAI 11326, pp. 199–212, 2019.
https://doi.org/10.1007/978-3-030-12738-1_15

and documenting goals of care and preferences for future care, is essential for achieving end-of-life care that is consistent with the preferences of seriously ill patients and their families. Inadequate communication is associated with more aggressive care near the time of death, decreased use of hospice and increased anxiety and depression in surviving family members [2–5]. Several studies have demonstrated the potential of advanced care planning to improve end-of-life outcomes (e.g., reducing unintended ICU admissions and increasing hospice enrollment). In the absence of explicit goals of care decisions, clinicians may provide clinical care [6] that does not provide a meaningful benefit to the patient [7] and, in the worse case, interferes with the treatment of other patients [6]. For these reasons, it is recommended that care preferences are discussed and documented in the EHR within the first 48 h of an ICU admission [8,9].

In recent years a consensus has emerged that such conversations are an essential component of practice and must be monitored to improve care quality. However, the difficulty of retrieving documentation about these conversations from the electronic health record has limited rigorous research on the prevalence and quality of clinical communication. For example, the National Quality Forum (NQF) recommends that goals of care be discussed and documented in the EHR within the first 48 h of an ICU admission, especially for frail and seriously ill patients. This was one of only two Centers for Medicare and Medicaid Services recommended palliative care quality measures for the Medicare Hospital Inpatient Quality Reporting program [10]. Yet, despite widespread support, routine assessment of this and similar quality measures have proven nearly impossible because the information is embedded as non-discrete free-text within clinical notes. Manual chart review is time-consuming and expensive to scale [11–13]. Consequently, many end-of-life quality metrics are simply not assessed, and their impact on distal and important patient outcomes have been insufficiently evaluated.

The emergence of omnipresent EHRs and powerful computers present novel opportunities to apply advanced computational methods such as natural language processing (NLP) [14] to assess end-of-life quality metrics including documentation of ACP. NLP enables machines to process or "understand" natural language in order to perform tasks like extracting communication quality embedded as non-discrete free-text within clinical notes [15].

Two main approaches to NLP information extraction exist. Rule-based extraction uses a pre-designed set of rules [14], which involves computing curated rules specified by experts, resulting in algorithms that detect specific words or phrases. This approach works well for smaller defined sets of data such as when searching for all the brand names of a generic medication (e.g., if X is present, then Y = 1). However, rule-based approaches fail when the desired information appears in a large variety of contexts within the free text [16].

Recent advances in machine learning coupled with increasingly powerful computers have created an opportunity to apply advanced computational methods, such as deep learning, to assess the content of free-text documentation within clinical notes. Such approaches possess the potential to broaden the scope of research on serious illness communication, and when implemented in real-time, to change clinical practice.

In contrast to rule-based methods, deep learning does not depend upon pre-defined set of rules. Instead, these algorithms learn patterns from a labeled set of free-text notes and apply them to future datasets [16]. A deep learning-based approach works well for tasks for which the set of extraction rules is very large, unknown, or both. In deep learning, algorithms can learn feature representations that aid in interpreting varied language.

In this study, we used deep learning [17] to train models to detect documentation of serious illness conversations, and we assess the performance of these deep learning models against manual chart review and rule based regular expression.

2   Data

2.1   Data Source

We derived our sample from the publicly available ICU database, Multi Parameter Intelligent Monitoring of Intensive Care (MIMIC) III, developed by the Massachusetts Institute of Technology (MIT) Lab for Computational Physiology and Beth Israel Deaconess Medical Center (BIDMC) [18]. It is a repository of de-identified administrative, clinical, and survival outcome data from more than 58,000 ICU admissions at BIDMC from 2001 through 2012. Between 2008 and 2012, the dataset also included clinical notes associated with each ICU admission. The Institutional Review Board of the BIDMC and MIT have approved the use of the MIMIC-III database by any investigator who fulfills data-user requirements. The study was deemed exempt by the Partners Institutional Review Board.

2.2   Cohort

The study population included adult patients (age ≥18) who were admitted to the medical, surgical, coronary care, or cardiac surgery ICU. The training and validation set included physician notes from patients who died during the hospital admission to ensure that we would have sufficient examples of documentation of care preferences. We excluded patients who did not have physician notes within the first 48 h because these patients either died shortly after admission or transferred out of the ICU.

2.3   Clinical Domains

Our main outcome was to identify documentation of care preferences within 48 h of an ICU admission in seriously ill patients. We aimed to detect the binary absence or presence of any clinical text that fit specified documentation of domains: patient care preferences (goals of care conversations or code status limitations), goals of care conversations, code status limitations, family communication (which included communication or attempt to communicate with family that did not result in documented care preferences), and full code status. Domains were chosen by board-certified, experienced palliative care clinicians through a lengthy and iterative process. They determined categories that are both relevant to widespread existing palliative care quality measures and interesting to future research questions. The specifications of each domain are outlined (Table 1).

Table 1. Clinical domain specifications.

Domain	Documentation example
Patient care preferences	Fulfills criteria for goals of care conversations and/or code status limitations
Goals of care conversations	Explicitly shown preferences about the patient's goals, values, or priorities for treatment and outcomes. Does NOT include presumed full code status or if obtained from other sources
Code status limitations	Explicitly shown preference of patient's care restricting the invasive care. Includes taken over preference from previous admission
Communication with family	Explicit conversations held during ICU stay period with patients or family members about the patient's goals, values, or priorities for treatment and outcomes
Full code status	Explicitly or implicitly shown preference for full set of invasive care including intubation and resuscitation. Includes presumed full code status or if obtained from other sources

2.4   Annotation

We developed a set of abstraction guidelines to ensure reliable abstraction between annotators. Each annotator identified clinical text that fit specified communication domains and labeled the portions of text identified for a domain, with no restrictions on length of a single annotation.

A gold standard dataset, considered to contain true positives and true negatives, was developed through manual annotation by a panel of four clinicians. Annotation was done using PyCCI, a clinical text annotation software developed by our team. Each note was annotated by at least two clinicians and annotations were then validated by a third clinician. Similar to previously published chart abstraction studies performed for this measure, the abstraction team had real-time access to a US board certified hospice and palliative medicine attending physician-expert reviewer, met weekly, and used a log to document common questions and answers to facilitate consistency [11,19].

The clinician coders manually annotated an average of 239 notes each (SD, 196), for a total of 641 notes. Each note contained an average of 1397 tokens (IQR, 1004-1710). The inter-rater reliability among the four clinician annotators was kappa >0.65 at the note level for each domain. The performance of each clinician coder was varied–for example, they identified documentation of care preferences with a sensitivity ranging from 77–92% (in comparison to the final gold standard).

3 Methods

3.1 Pre-processing

Annotated notes were pre-processed for both rule-based regular expression and neural network methods. First, texts were cleaned to remove any extraneous spaces, lines, or characters. Each cleaned note was tokenized, which means it was split into identifiable elements–in this case, words and punctuation. We used the Python module spaCy in order to tokenize intelligently, based on the structure of the English language [20]. Labels were associated with individual tokens and datasets were split out by domain, as each method was run separately.

3.2 Regular Expression

Our baseline model is a simple regular expression based on pre-curated rules for each domain. Appendix A shows the rules used for each domain. These rules are keywords that the regular expression program identifies as belonging to its corresponding domain, taking into account varieties in punctuation and case. To create the regular expression library, we identified tokens that were sensitive and specific for each prediction task. We calculated sensitivity by evaluating the proportion of a token's total number of occurrences that were labeled for each domain. We evaluated specificity by evaluating what proportion of a token's total number of occurrences were in a note that was in an unlabeled note for each domain. A board-certified clinician used these data points–sensitivity, specificity, frequency that each token appeared on the labeled text and frequency in texts outside of the domain–and their clinical knowledge to generate a list of terms that could likely be generalized.

Regular expressions identify patterns of characters exactly as they are specified in a set of rules. If text in the note matches a keyword in the regular expression library for the domain, it is labelled as positive for that concept. This method acts as a baseline to compare our algorithm against. We used a regular expression program, ClinicalRegex, also developed by our lab [30]. ClinicalRegex is easily accessible and intuitive to navigate, which makes it an efficient choice for groups that are not able to employ computer scientists. We have chosen to compare our deep learning methods against an easily understandable and accessible method to illustrate the benefits of more complex methods.

3.3 Artificial Neural Network

Deep learning involves training a neural network to learn data representation and fulfill a specified task. We trained algorithms to identify clinical text documentation of serious illness communication. During the training process, the neural network learns to identify and categorize tokens (individual words and symbols) as belonging to each of the pre-specified domains and maximizes probability across predicted token labels [21].

The specific neural network used, NeuroNER, was developed by Dernoncourt et al. for the purpose of named-entity recognition [22]. NeuroNER has been evaluated for use in the de-identification of patient notes [21]. It allows for each token to be labelled only with a single label. However, tokens in our study were often associated with multiple labels. For example, a sentence could indicate that both communication with family occurred and that goals of care were discussed. In order to allow for multi-class labelling, a separate, independent model was trained per domain. For each domain, the data set was split up into randomized training and validation sets, with 70% (449 notes) of the set in training, and 30% (192 notes) in validation.

With the parameters derived from this training process, the model is run on the validation data set to examine its performance on a data set it was not specifically tuned to fit. Performance on the validation set also determines when training converges, indicating that the model is optimally trained. Training converges when there has been no improvement on the validation set performance in ten epochs. The neural network ultimately determines domain labels for each token. From the predicted token-level results, a note-level classification is determined by the presence or absence of labelled tokens by domain in each note. We used Tensorflow version 1.4.1 and trained our models on a NVIDIA Titan X Pascal GPU. Below are the hyperparameters selected for our use:

- character embedding dimension: 25
- character-based token embedding LSTM dimension: 25
- token embedding dimension: 100
- label prediction LSTM dimension: 100
- dropout probability: 0.5.

For our experiments, we were able to compare our gold standard labels, derived from manual annotation by clinicians as described in Sect. 2.4, to the predicted output to evaluate the performance of the neural network and the regular expression method.

4   Results

4.1   Evaluation Metrics

Algorithm performance was determined at two levels: token-level and note-level, referring to the binary absence or presence of a label at these levels. Token-level results are more specific and allow accurate identification of relevant text within clinical notes. Note-level results allow determination of whether documentation of communication occurred. At both of these levels, we calculated the following metrics: sensitivity, specificity, positive predictive value, accuracy, and F1-score. The F1-score is the harmonic average of positive predictive value and sensitivity. It allows us to determine the success of our algorithm both in identifying true positives as well as true negatives.

The 95% confidence intervals for all metrics were determined via bootstrapping [23]; each trained network model was validated for 1,000 trials in addition to the reported performance point. During each trial, a validation set of 192 notes was created by random sampling with replacement of the original validation set of 192 unique notes. This creates an approximate distribution of performance for the model. In basic bootstrap technique, the 2.5th and 97.5th percentiles of the distributions for each metric are taken as the 95% confidence interval [24].

4.2   Performance

Table 2 summarizes the performance of the regular expression method and Table 3 summarizes the performance of the neural networks in identifying documentation of serious illness communication at the note level, for each clinical domain, on the validation set. Figure 1 displays a comparison in the F1-scores for each domain. For identification of documentation of patient care preferences, the algorithm achieved an F1-score of 92.0 (95% CI, 89.1–95.1), with 93.5% (95% CI, 90.0%–98.0%) sensitivity, 90.5% (95% CI, 86.4%–95.1%) positive predictive value and 91.0% (95% CI, 86.4%–95.3%) specificity. For identification of family communication without documentation of preferences, the algorithm achieved an F1-score of 0.91 (95% CI, 0.87–0.94), with 90.7% (95% CI, 86.0%–95.9%) sensitivity, 90.7% (95% CI, 86.5%–94.8%) positive predictive value and 92.5% (95% CI, 89.2%–97.8%) specificity. Token-level performance is displayed in Appendix B.

At the note-level, we have been able to achieve high accuracy for all domains and see that in the validation set, the neural network outperforms the regular expression method in every domain for F1-score, significantly so in identifying patient care preferences, goals of care conversations, and communication with family. These domains contain more complex and diverse language, which are successfully identified by the neural network. A static library is not able to capture the diversity in these domains, necessitating the use of machine learning.

4.3   Error Analysis

A review of documentation that the neural networks identified as serious illness conversations that was not labeled serious illness conversations in the gold standard (false positives) showed that the algorithm identified documentation that clinician coders missed. Though our gold standard was rigorously reviewed and validated, there still remains room for human error. Comparing the identified text from the neural network and regular expression methods, we found that

Table 2. Performance (%) of the regular expression method on the validation data set.

Domain	F1-score	Accuracy	Sensitivity	Positive predictive value	Specificity
Patient care preferences	76.0	78.6	70.7	82.3	86.0
Goals of care conversations	37.2	57.8	26.1	64.9	87.0
Code status limitations	94.3	96.4	98.3	90.6	95.5
Communication with family	43.6	67.7	27.9	100.0	100.0
Full code status	90.9	88.5	84.6	98.2	96.8

Table 3. Performance (%) of the neural networks on the validation data set. Values in parentheses are 95% confidence intervals.

Domain	F1-score	Accuracy	Sensitivity	Positive predictive value	Specificity
Patient care preferences	92.0 (89.1–95.1)	92.2 (89.6–95.1)	93.5 (90.0–98.0)	90.5 (86.4–95.1)	91.0 (86.4–95.3)
Goals of care conversations	85.7 (80.4–90.3)	89.1 (85.6–92.4)	85.1 (78.4–91.5)	86.3 (80.0–93.0)	91.5 (87.7–95.7)
Code status limitations	95.9 (93.0–98.7)	97.4 (95.8–99.2)	98.3 (96.9–100.0)	93.5 (89.2–97.7)	97.0 (95.0–98.9)
Communication with family	90.7 (87.4–93.9)	91.7 (89.1–94.4)	90.7 (86.0–95.9)	90.7 (86.5–94.8)	92.5 (89.1–95.9)
Full code status	98.5 (97.5–99.4)	97.9 (96.6–99.2)	100.0 (100.0–100.0)	97.0 (95.1–98.9)	93.5 (89.2–97.7)

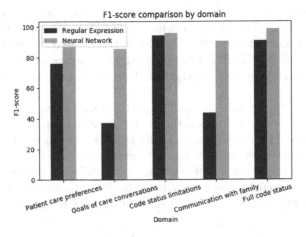

Fig. 1. Comparison between the F1-score of the regular expression method and neural networks by domain.

as expected, the neural network was able to identify complex and unique language that the regular expression method was not. Doctors employ diverse and non-standardized language in clinical notes; we require more flexible and extensible methods in order to efficiently process this information. Static libraries cannot capture the full complexity of language without sacrificing sensitivity or specificity–they must be curated such that library terms are not too broad and they are not able to utilize context. All note-level identification can be traced to the detection of specific words with examples of text for each method provided in Appendix C.

4.4   Effect of Training Set Size

In order to determine how smaller training sets related to the performance of the trained algorithms, we trained multiple networks with varying number of notes. We plotted training dataset size against algorithm performance for 8 sample sizes (Fig. 2). The performance seemed to plateau at around 200 notes (around 250,000 tokens), which suggests that annotation efforts can be efficiently leveraged to generalize the models to varied health systems.

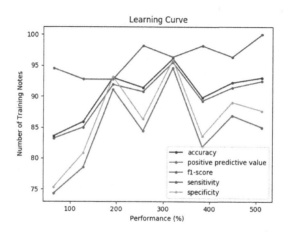

Fig. 2. Neural network performance on validation set for detection of note-level documentation of patient care preferences by number of notes used for training.

5   Discussion and Future Work

We describe a novel use of deep learning algorithms to rapidly and accurately identify documentation of serious illness conversations within clinical notes. When applied to identifying documentation of patient care preferences, our algorithm demonstrated high sensitivity (93.5%), positive predictive value (90.5%) and specificity (91.0%), with a F1-score of 92.0. In fact, we found that deep learning outperformed individual clinician coders both in terms of identifying the documentation and in terms of its many-thousands-time-faster speed.

Existing work has shown that machine learning can extract structured entities like medical problems, tests and treatments from clinical notes [25,26], and unstructured image-based information in radiology, pathology and opthamology [27–29]. Our study extends this line of work and demonstrates that deep learning can also perform accurate automated text-based information classification.

Up until now, extracting goals of care documentation nested within free-text clinical notes has relied on labor-intensive and imperfect manual coding [11]. Using the capabilities of deep learning as demonstrated in this paper would allow for rapid audit and feedback regarding documentation at the system and individual practitioner level. This would result in significant opportunities for quality improvement that are currently not being met. Deep learning models could also improve patient care in real-time by broadening what is available at the point of care in the EHR. For example, clinicians could view displays of all documented goals of care conversations, or be prompted to complete documentation that was not yet available.

Important limitations must be noted. Deep learning algorithms only detect what is documented. It is not fully understood to what extent documentation reflects the actual content of a patient-clinician conversation surrounding serious illness care goals. However, documentation is the best proxy we have to understand and to track these conversations. This is also a single institution study, which may limit its generalizability. Future work will involve the investigation of how extensible models are to clinical notes from different health system. Variations in EHR software and the structure of clinical notes in different institutions makes it essential to further train and validate our methods using data from multiple healthcare systems. This should be imminently possible, as our learning curve suggested that the neural network needed to train on as few as 200 clinician coded notes to perform well. Future research should also focus on optimizing deep neural networks to further improve performance, and on determining the feasibility of operationalizing this algorithm across institutions.

6   Conclusion

This is the first known report of employing deep learning, to our knowledge, to identify serious illness conversations. The potential of this technology to improve the visibility of documented goals of care conversations within the EHR and for quality improvement has far reaching implications. We hope such methods will become an important tool for evaluating and improving the quality of serious illness care from a population health perspective.

Acknowledgements. We are particularly grateful to Tristan Naumann, Franck Dernoncourt, Elena Sergeeva, Edward Moseley, and Alistair Johnson for helpful guidance and advice during the development of this research. Additionally, we would like to thank Peter Szolovits for providing computing resources, as well as Saad Salman, Sarah Kaminar Bourland, Haruki Matsumoto and Dickson Lui for annotating clinical notes. This research was facilitated by preliminary work done as part of course HST.953 in the Harvard-MIT Division of Health Sciences and Technology (HST) at Massachusetts Institute of Technology (MIT), Boston, MA.

A   Regular Expression Library

Domain	Keywords
Patient care preferences	goc, goals of care, goals for care, goals of treatment, goals for treatment, treatment goals, family meeting, family discussion, family discussions, patient goals, dnr, dni, dnrdni, dnr/dni, DNI/R, do not resuscitate, do-not-resuscitate, do not intubate, do-not-intubate, chest compressions, no defibrillation, no endotracheal intubation, no mechanical intubation, shocks, cmo, comfort measures
Goals of care conversations	goc, goals of care, goals for care, goals of treatment, goals for treatment, treatment goals, family meeting, family discussion, family discussions, patient goals
Code status limitations	dnr, dni, dnrdni, dnrdni, DNIR, do not resuscitate, do-not-resuscitate, do not intubate, do-not-intubate, chest compressions, no defibrillation, no endotracheal intubation, no mechanical intubation, shocks, cmo, comfort measures
Communication with family	Explicit conversations held during ICU stay period with patients or family members about the patient's goals, values, or priorities for treatment and outcomes
Full code status	full code

B   Token-Level Performance

See Table 4.

Table 4. Performance (%) of the neural network on the validation data set at the token-level.

Domain	F1-score	Accuracy	Sensitivity	Positive predictive value	Specificity
Patient care preferences	76.0	99.6	75.8	75.2	99.8
Goals of care conversations	70.4	99.6	70.0	69.9	99.8
Code status limitations	76.3	99.8	72.7	80.5	99.9
Communication with family	68.2	99.7	62.0	76.4	99.9
Full code status	90.9	99.8	88.3	93.6	99.8

C   Examples of Identified Text

Below are examples of correctly identified serious illness documentation by the neural network and regular expression methods in the validation dataset. Correctly identified tokens are bolded. Typographical errors are from the original text. Each cell includes an example of identified tokens in the same text and an example of documentation identified by the neural network that was missed by the regular expression method, if relevant.

Domain	Neural network	Regular expression
Goals of care conversations	Hypercarbic resp failure: family meeting was held with son/**HCP and in keeping with patients goals of care, there was no plan for intubation.** Family was brought in and we explained the graveness of her ABG and her worsened mental status which had failed to improve with BiPAP. **Family was comfortable with removing Bipap and providing comfort care** including morphine prn **family open to cmo but pt wants full treatment or to be disturbed**	Hypercarbic resp failure: **family meeting** was held with son/HCP and in keeping with patients **goals of care**, there was no plan for intubation.Family was brought in and we explained the graveness of her ABG and her worsened mental status which had failed to improve with BiPAP. Family was comfortable with removing Bipap and providing comfort care including morphine prn family open to cmo but pt wants full code but also doesn't want treatment or to be disturbed
Code status limitations	**CODE: DNR/DNI, confirmed with healthcare manager who will be discussing with official HCP**	CODE: **DNR/DNI**, confirmed with healthcare manager who will be discussing with official HCP
Communication with family	Dr. [**First Name (STitle) **] from **neurosurgery held family meeting and explained grave prognosis to the family** **lengthy discussion with the son who is health care proxy he wishes to pursue comfort measures due to severe and unrevascularizable cad daughter is not in agreement at this time but is not the proxy due to underlying psychiatric illness**	Dr. [**First Name (STitle) **] from neurosurgery held **family meeting** and explained grave prognosis to the family lengthy discussion with the son who is health care proxy he wishes to pursue comfort measures due to severe and unrevascularizable cad daughter is not in agreement at this time but is not the proxy due to underlying psychiatric illness
Full code status	**Code: FULL; Discussed with daughter and HCP who says that patient is in a Hospice program with a "bridge" to DNR/DNI/CMO, but despite multiple conversations, the patient insists on being full code** **CODE: Presumed full**	Code: FULL; Discussed with daughter and HCP who says that patient is in a Hospice program with a "bridge" to DNR/DNI/CMO, but despite multiple conversations, the patient insists on being **full code** CODE: Presumed full

References

1. Cook, D., Rocker, G.: Dying with dignity in the intensive care unit. New Engl. J. Med. **370**, 2506–2514 (2014)
2. Wright, A.A., Zhang, B., Ray, A., et al.: Associations between end-of-life discussions, patient mental health, medical care near death, and caregiver bereavement adjustment. JAMA **300**(14), 1665–1673 (2008)
3. Nicholas, L.H., Langa, K.M., Iwashyna, T.J., Weir, D.R.: Regional variation in the association between advance directives and end-of-life Medicare expenditures. JAMA **306**(13), 1447–1453 (2011)
4. Teno, J.M., Gruneir, A., Schwartz, Z., Nanda, A., Wetle, T.: Association between advance directives and quality of end-of-life care: a national study. J. Am. Geriatr. Soc. **55**(2), 189–194 (2007)
5. Detering, K.M., Hancock, A.D., Reade, M.C., Silvester, W.: The impact of advance care planning on end of life care in elderly patients: randomised controlled trial. BMJ **340**, c1345 (2010)
6. Huynh, T.N., Kleerup, E.C., Raj, P.P., Wenger, N.S.: The opportunity cost of futile treatment in the intensive care unit. Crit. Care Med. **42**(9), 1977–1982 (2014). https://doi.org/10.1097/CCM.0000000000000402
7. Huynh, T.N., et al.: The frequency and cost of treatment perceived to be futile in critical care. JAMA Intern. Med. **173**, 1887–1994 (2013)
8. NQF #1626: Patients Admitted to ICU Who Have Care Preferences Documented. National Quality Forum
9. Khandelwal, N., Kross, E., Engelberg, R., Coe, N., Long, A., Curtis, J.: Estimating the effect of palliative care interventions and advance care planning on ICU utilization: a systematic review. Crit Care Med. **43**, 1102–1111 (2015). https://doi.org/10.1097/CCM.0000000000000852
10. Rising, J., Corrigan, J., Valuck, T.: Building Additional Serious Illness Measures Into Medicare Programs. The Pew Charitable Trusts, Philadelphia (2017)
11. Walling, A.M., Tisnado, D., Asch, S.M., et al.: The quality of supportive cancer care in the veterans affairs health system and targets for improvement. JAMA Intern. Med. **173**(22), 2071–2079 (2013)
12. Dy, S.M., Lorenz, K.A., O'Neill, S.M., et al.: Cancer quality-ASSIST supportive oncology quality indicator set: feasibility, reliability, and validity testing. Cancer **116**(13), 3267–3275 (2010)
13. Aldridge, M.D., Meier, D.E.: It is possible: quality measurement during serious illness. JAMA Intern. Med. **173**(22), 2080–2081 (2013)
14. Melton, G.B., Hripcsak, G.: Automated detection of adverse events using natural language processing of discharge summaries. JAMA **12**(4), 448–457 (2005)
15. Honnibal, M., Johnson, M.: An improved non-monotonic transition system for dependency parsing. In: Proceedings of the 2015 Conference on Empirical Methods in Natural Language Processing, September 2015, Lisbon, Portugal, pp. 1373–1378. Association for Computational Linguistics (2015)
16. Carrell, D.S., et al.: Challenges in adapting existing clinical natural language processing systems to multiple, diverse healthcare settings. (JAMIA) **2**, 986–991 (2017)
17. Schmidhuber, J.: Deep learning in neural networks: an overview. Neural Netw. **61**, 85–117 (2015). arXiv:1404.7828. https://doi.org/10.1016/j.neunet.2014.09.003. PMID 25462637
18. Johnson, A.E., et al.: MIMIC-III, a freely accessible critical care database. Sci. Data **3**, 160035 (2016)

19. Walling, A.M., et al.: The quality of care provided to hospitalized patients at the end of life. Arch. Intern. Med. **170**(12), 1057–1063 (2010)

20. Honnibal, M., Johnson, M.: An improved non-monotonic transition system for dependency parsing. In: Proceedings of the 2015 Conference on Empirical Methods in Natural Language Processing, Lisbon, Portugal (2015)

21. Dernoncourt, F., Lee, J.Y., Uzuner, O., Szolovits, P.: De-identification of patient notes with recurrent neural networks. J. Am. Med. Inform. Assoc. **24**(3), 596–606 (2017)

22. Dernoncourt, F., Lee, J.Y., Szolovits, P.: NeuroNER: an easy-to-use program for named-entity recognition based on neural networks. In: Conference on Empirical Methods on Natural Language Processing (EMNLP) (2017)

23. Efron, B.: Better bootstrap confidence intervals. J. Am. Stat. Assoc. **82**(397), 171–185 (1987)

24. Davison, A.C., Hinkley, D.V.: Bootstrap Methods and their Application. Cambridge University Press, Cambridge (1997)

25. D'Avolio, L.W., Nguyen, T.M., Goryachev, S., Fiore, L.D.: Automated concept-level information extraction to reduce the need for custom software and rules development. J. Am. Med. Inform. Assoc. **18**(5), 607–613 (2011)

26. Xu, H., et al.: Facilitating pharmacogenetic studies using electronic health records and natural-language processing: a case study of warfarin. J. Am. Med. Inform. Assoc. **18**(4), 387–391 (2011)

27. Bejnordi, B.E., Veta, M., van Diest, P.J., et al.: Diagnostic assessment of deep learning algorithms for detection of lymph node metastases in women with breast cancer. J. Am. Med. Inform. Assoc. **318**(22), 2199–2210 (2017)

28. Gulshan, V., Peng, L., Coram, M., et al.: Development and validation of a deep learning algorithm for detection of diabetic retinopathy in retinal fundus photographs. J. Am. Med. Inform. Assoc. **316**(22), 2402–2410 (2016)

29. Ting, D.S.W., et al.: Development and validation of a deep learning system for diabetic retinopathy and related eye diseases using retinal images from multiethnic populations with diabetes. J. Am. Med. Inform. Assoc. **318**, 2211–2223 (2017)

30. Lindvall, C., et al.: Natural Language Processing to Assess End-of-Life Quality Indicators in Cancer Patients Receiving Palliative Surgery. J Palliatd Med., 17 October 2018. https://doi.org/10.1089/jpm.2018.0326

Generating Reward Functions Using IRL Towards Individualized Cancer Screening

Panayiotis Petousis[1]([✉]) [iD], Simon X. Han[1] [iD], William Hsu[1,2] [iD], and Alex A. T. Bui[1,2] [iD]

[1] UCLA Bioengineering Department, Los Angeles, CA 90095, USA
pp89@ucla.edu
[2] UCLA Department of Radiological Sciences, Los Angeles, CA 90095, USA

Abstract. Cancer screening can benefit from individualized decision-making tools that decrease overdiagnosis. The heterogeneity of cancer screening participants advocates the need for more personalized methods. Partially observable Markov decision processes (POMDPs), when defined with an appropriate reward function, can be used to suggest optimal, individualized screening policies. However, determining an appropriate reward function can be challenging. Here, we propose the use of inverse reinforcement learning (IRL) to form rewards functions for lung and breast cancer screening POMDPs. Using experts (physicians) retrospective screening decisions for lung and breast cancer screening, we developed two POMDP models with corresponding reward functions. Specifically, the maximum entropy (MaxEnt) IRL algorithm with an adaptive step size was employed to learn rewards more efficiently; and combined with a multiplicative model to learn state-action pair rewards for a POMDP. The POMDP screening models were evaluated based on their ability to recommend appropriate screening decisions before the diagnosis of cancer. The reward functions learned with the MaxEnt IRL algorithm, when combined with POMDP models in lung and breast cancer screening, demonstrate performance comparable to experts. The Cohen's Kappa score of agreement between the POMDPs and physicians' predictions was high in breast cancer and had a decreasing trend in lung cancer.

Keywords: Cancer screening ·
Maximum entropy inverse reinforcement learning ·
Partially-observable Markov decision processes

1 Introduction

Annually, millions of people undergo screening for disease prevention and surveillance. From these tests, physicians aim to make decisions based on the patient's past results and most current observations, determining a subsequent action (e.g., further diagnostic testing, increased monitoring, following regular screening

© Springer Nature Switzerland AG 2019
F. Koch et al. (Eds.): AIH 2018, LNAI 11326, pp. 213–227, 2019.
https://doi.org/10.1007/978-3-030-12738-1_16

schedules, etc.) that optimizes early detection of health problems while balancing other (pragmatic) concerns (e.g., patient quality of life, resource utilization, cost). Choosing the "best" next step and tailoring screening for each person is challenging: selecting an action of benefit in the immediate future may not be optimal over the long-term, given the particulars of an individual (i.e., a locally greedy approach vs. a global optimization).

Sequential decision making methods provide a potential solution. Such approaches can integrate and analyze multiple sources of patient data, while handling issues related to temporal credit assignment. In particular, partially observable Markov decision processes (POMDPs) have been applied to cancer screening (e.g., breast, colorectal, prostate [20]) to determine policies based on patients' risk factors and prior screening results. Markedly, POMDP models used in medicine typically use a reward function adopted from cost-effectiveness studies [20] or are posed in terms of quality-adjusted life years (QALYs). While such functions are informative about general populations, they do not necessarily reflect how an experienced clinician would make a decision, especially given a specific individual's medical history and preferences. Indeed, little work has been done in designing reward functions that emulate experts' decision processes.

Here, we propose using the Maximum Entropy Inverse Reinforcement Learning (MaxEnt IRL) algorithm [26] to establish reward functions from retrospective screening data, learning how an expert physician may select a given action based on observed test results. We use an adaptive step size to expedite the convergence rate of MaxEnt IRL. Importantly, we present how to use the MaxEnt IRL learned rewards to generate state-action pair rewards that can be used in POMDPs. We demonstrate this work using two real-world clinical datasets for lung and breast cancer screening, mimicking how clinicians made decisions regarding patients. We evaluate the resultant POMDP policies using the MaxEnt IRL reward functions, comparing model performance to experts' actions. We conclude that the MaxEnt IRL algorithm is an efficient and accurate method in estimating sensible reward functions for cancer screening.

2   Background

Although Markov decision processes (MDPs) and POMDPs are used in a number of domains, their application in healthcare is limited and few strategies exist for estimating the associated reward functions that drive agent behavior in clinical settings. Taken from the perspective of epidemiological and health services research, different cost and patient benefit metrics are frequently adapted for optimization. Classic examples include: Bennet et al. [5], who proposed a cost-effectiveness metric based on the cost required to obtain one unit of outcome change (CPUC); Hauskrecht et al. [12], who designed a reward model that combines economic cost and patient quality of life measures; and Tusch et al. [22], who predicated rewards on 30-day mortality risk for a surgical procedure. In contrast, we take advantage of growing amounts of longitudinal data, using recorded information and actions from electronic health records (EHRs) and other observational data sources, to learn a POMDP reward function that imitates expert physicians' behavior for desired health outcomes. Specifically, IRL is proposed for this task.

Briefly, IRL addresses the problem of obtaining a reward function given an agent's optimal behavior over time towards a stated goal. A reward function for the environment is unknown and is hence learned through empirical investigation of sensory inputs (i.e., observations) that progressively change the agent's selection of different actions. Two families of IRL algorithms exist: (1) linear programming (LP) methods [1,18]; and (2) probabilistic IRL algorithms [4,26]. While potentially more computationally complex, probabilistic IRL approaches have two advantages: they guarantee a unique solution for deterministic MDPs; and compared to LP methods, they can handle stochasticity in the data [23]. Vroman et al. [4] developed a maximum likelihood IRL algorithm using clusters of experts' data trajectories to characterize different intentions. Applying the maximum likelihood IRL algorithm to each cluster subsequently derives a reward function representing the experts' behavior. Ziebart et al. [25,26] describe a probabilistic IRL algorithm that employs the principle of maximum entropy, dealing with noise and imperfect behavior as it normalizes globally over behaviors. In this approach, demonstrated for modeling routing preferences of vehicle drivers, behaviors with higher rewards are exponentially preferred by the algorithm when learning the reward function. Here, we build on and adapt this approach to obtain reward functions for cancer screening POMDPs.

3   Materials and Methods

3.1   NLST Dataset

The National Lung Screening Trial (NLST) is a multi-site randomized controlled trial that demonstrated a 20% mortality reduction in lung cancer screening using low-dose computed tomography (LDCT) relative to plain chest radiography [17]. For this work, we used data from the NLST's LDCT arm, comprising approximately 25,500 participants that underwent three annual screenings and follow-up post screening. We further filter this dataset to those subjects who had a reported pulmonary nodule based on imaging. Unfortunately, preprocessing of the NLST data is not straightforward, as longitudinal tracking of the nodules was not considered at the time of the study. Thus, to use imaging-related information, we made the assumption that an imaging finding in individuals with only one reported nodule and in the same anatomical location over time is the same nodule across the three screening points of the trial. This criterion further constrained our dataset to 5,402 LDCT subjects. From this subgroup, we learned a reward function, then trained and tested a POMDP. Note that for the reward function we made use of the recorded diagnostic follow-up variables (e.g., recommendation for other procedures) to inform actions.

3.2   Athena Dataset

The Athena Breast Health Network [10] is a University of California (UC)-wide initiative around breast cancer screening and treatment. The effort started in

2009 and includes women who underwent breast screening at five academic medical centers. The portion available at our institution (UCLA) consists of 49,244 patients, with follow-ups of up to 4.8 years; this subset represents 96,515 screening and diagnostic mammograms (MGs), and 2,713 diagnostic biopsies. MG results are reported as Breast Imaging Reporting and Data System (BI-RADS) scores [9]. We selected patients with initial risk (Gail) scores, four consecutive screenings, valid BI-RADS scores, and biopsies results per breast side (i.e., left, right). 2,095 patients with left breast MGs and 2,036 patients with right breast MGs (4,131 total cases, 4,099 after pre-processing) were used in this study.

3.3   Partially Observable Markov Decision Processes

An MDP is represented by a tuple of states, actions, rewards, action-dependent state transition dynamics (i.e., transition probabilities), and a discount factor. A POMDP is an extension to MDPs with two additional components: observations and state-dependent observation dynamics (i.e., observation probabilities). The state of the agent in POMDPs is partially observable. As such, its state is modeled as a probability distribution over the states, called the belief state, which is updated over time based on the observations experienced by the agent.

We designed and evaluated two separate POMDPs for lung and breast cancer screening. Each model consists of three states and two actions. The observations of each POMDP are domain based: in the lung model, they represent findings obtained from LDCT imaging studies, including nodule size, consistency, location, and margins; in the breast model, they represent BI-RADS scores derived from MG interpretations. Given the nature of each dataset, both the lung and breast models have a horizon of three and four years, respectively, with 6-month and 1-year epochs. Each epoch represents time points for which we have information on the cancer status of patient (diagnosed with cancer or not). Transition and observation probabilities for each POMDP model are learned using the expectation maximization (EM) algorithm, for learning dynamic Bayesian networks, from each dataset. Both models were solved using the QMDP approximation solver [21].

Lung Cancer Screening POMDP. Figure 1 (left) depicts the lung POMDP, illustrating the state space and allowed transitions between states, as well as the observations of each state. The state space consists of three states: the no-cancer (NC) state that represents any case with no suspicious abnormalities (i.e., no pulmonary nodules >4 mm). The uncertain (U) state that represents any case with a noted finding (i.e., nodules 4 mm or larger) but not yet a lung cancer. Lastly, the invasive-cancer (IC) state is any case with a confirmed lung cancer diagnosis through the use of additional diagnostic tests. The IC state is terminal such that any individual who enters it leaves the screening process for treatment. An LDCT action implies continuation of screening, whereas an intervention action refers to any diagnostic procedure (e.g., thoracotomy, biopsies, diagnostic CT,

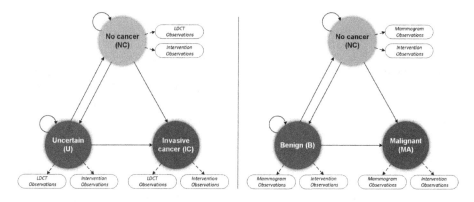

Fig. 1. Left. The lung POMDP; NC: no-cancer state; U: uncertain state; IC: invasive cancer state. LDCT and intervention observations can be observed in each state. **Right**. The breast POMDP; NC: non-cancer state; B: benign state; MA: malignant cancer state. MG and intervention observations can be observed in each state.

positron emissions tomography (PET) scan). Observations represent LDCT findings (nodule size, consistency, margins, and anatomic location) and the occurrence of an intervention. To generate initial belief states for each individual in our dataset we used the Tammemägi PLCO$_{M2012}$ model with demographic and clinical features at baseline to predict the risk of cancer. Demographic features used include age, education, race, and body mass index. Clinical features used were COPD, family history of lung cancer, personal history of cancer, smoking status, smoking intensity, and duration of smoking.

Breast Cancer Screening POMDP. The breast POMDP model also consists of three states: the no-cancer (NC) state in which no abnormalities are seen, the benign (B) state in which benign breast disease diagnosis follows the MG, and the malignant (MA) cancer state in which the disease is confirmed through biopsy. MA is similarly a terminal state in which the patient leaves the screening process for treatment. Figure 1 (right) shows the breast cancer screening POMDP, transitions, observations (BI-RADS scores 1, 2, 3, 4A, 4B, 4C, 5), and actions. Though an intervention (biopsy in the breast cancer context) is possible after each MG, in practice biopsies are only performed after an MG of BI-RADS 4 or higher. For an initial belief, we used the patient's Gail score. The Gail score is an absolute risk estimate derived using age, age at menarche, age at first birth, the number of first-degree relatives with breast cancer, the number of previous breast biopsies, and race.

3.4 Maximum Entropy IRL

In IRL, the reward function, r, is assumed to be a linear combination of feature vectors f_s and weights θ (θ^T is the transpose of θ):

$$r(\tau;\theta) = \theta^T f_\tau = \sum_{s \in \tau} \theta^T f_s \tag{1}$$

A feature count, (f_τ), is the sum of feature vectors of the states visited along a trajectory, where f_s represents binary vectors indicating state values. Inputs to the MaxEnt IRL algorithm are an MDP and a set of trajectories (D) [3]. A path or a trajectory (τ) represents the sequence of states (s) and ensuing actions followed by an agent in an MDP. For example, in the NLST dataset, a trajectory comprises three epochs (i.e., the three annual screening exams) with state-action pairs describing the lung cancer states and the actions taken (e.g., NC-LDCT, U-LDCT, and IC-I$_{\text{Biopsy}}$). The probability of a trajectory occurring in our set of trajectories is proportional to the exponential of the reward/cost of the trajectory [7]:

$$p(\tau;\theta) \propto \exp\left(r(\tau;\theta)\right) \tag{2}$$

As such, trajectories of equal reward are equally likely to be executed by the expert, whereas trajectories of less reward are less likely. The probability distribution over paths with maximum information entropy is parameterized over θ. $Z(\theta)$ is the partition function, where $Z(\theta) = \sum_{\tau \in D} \exp r(\tau;\theta)$.

$$p(\tau;\theta) = \frac{1}{Z(\theta)} \exp\left(r(\tau;\theta)\right) \tag{3}$$

The log likelihood of the trajectories (loss function) is shown in Eq. 4, M is the number of trajectories:

$$L = \frac{1}{M} \sum_{\tau \in D} r(\tau;\theta) - \log \sum_{\tau \in D} \exp\left(r(\tau;\theta)\right) \tag{4}$$

This loss function is convex for a linear reward function and a deterministic MDP. To update θ we use a gradient descent function, where η represents the learning rate:

$$\theta_{i+1} = \theta_i + \eta \nabla_\theta L \tag{5}$$

The gradient $\nabla_\theta L$ represents the difference of feature expectations and sum over state visitation frequencies multiplied with feature vectors:

$$\nabla_\theta L = \tilde{f} - \sum_{s_i} D_{s_i} f_{s_i} \tag{6}$$

A feature expectation, $(\tilde{f})$, is defined as the average of all feature counts across all trajectories. The frequency of state visitation, D_{s_i}, can be computed using a dynamic programming algorithm; see [3,7] for more information regarding this algorithm. The pseudocode of the MaxEnt IRL algorithm can be found in [7].

3.5   Adaptive Step Size

To improve the convergence of the MaxEnt IRL algorithm, we introduce an adaptive learning rate approach for the update rule of the gradient descent. The idea behind making the step size adaptive is to calculate the inner product of $\nabla_\theta L$, the gradient, in the current step, i.e., $\nabla_\theta L_i$ with $\nabla_\theta L_{i-1}$, its value from the previous step. If the two are in the same direction then the step size can be increased, otherwise it is decreased. Following [15] we define the learning rate $\eta = \frac{\alpha}{(t+A)^\alpha}$, where t is dependent on the gradient inner product (which becomes the dot product in higher dimensions); α and A are constants. The role of t is to regulate the learning rate:

$$t_{i+1} = \max(t_i + f(\langle -\nabla_\theta L_i, \nabla_\theta L_{i-1}\rangle), 0) \qquad (7)$$

In this definition, $f(\cdot)$ represents the following sigmoidal function where $f(x) = f_{min} + \frac{f_{max}-f_{min}}{1-\frac{f_{max}}{f_{min}}\exp-\frac{x}{\omega}}$. In the above expressions, α, A, f_{min}, f_{max}, and ω are user-defined constants obtained from [15]. With $f_{min} < 0$, $f_{max} > 0$, and $\omega > 0$.

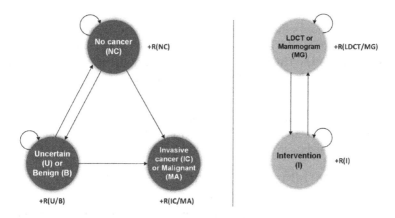

Fig. 2. Left. The state MDP; NC: non-cancer state; U/B: uncertain or benign state; I/MA: invasive or malignant cancer state, respectively for the lung and breast models. **Right**. The action MDP; LDCT/MG: state after a LDCT or MG; I: state after an intervention (e.g., biopsy); +R(·): rewards experienced by the agent in each state.

3.6   Computation of Rewards

We assumed that given the outcome of a known cancer diagnosis for each individual over time, partial observability was no longer a problem while training, so learning the rewards of state-action pairs of an MDP instead of a POMDP was sufficient and computationally more efficient. However, the MaxEnt IRL algorithm computes the rewards of each state of an MDP, not state-action pair rewards $(r(s,a))$. To estimate rewards for each state-action pair combination, we designed two MDPs:

1. *A state MDP model.* The states of this MDP are the states depicted in Fig. 2, for the lung and breast models. The transition matrix of the state MDP is the same transition matrix used in its respective POMDP model.
2. *An action MDP model.* In the action MDP, the states are defined by the previous action of the agent. These states model the options for screening (e.g., continue annual screening) and intervention (e.g., biopsy), in which the agent enters after performing each action. The action MDP transition model represents the probability of transitioning from the LDCT/MG state to the I state.

Figure 2 demonstrates the two MDPs. A combinatorial design decision inspired by [13] was used to learn state-action pair rewards. State-action pair rewards are computed using a multiplicative model shown in Eq. 8:

$$R(s, a) = R(s) \cdot R(a) \tag{8}$$

4   Evaluation and Results

A stratified 5-fold cross validation study design was used to evaluate the POMDP models built from the NLST and the Athena datasets. The training set of each fold is used to learn the transition and observation matrices of the POMDPs, as well as the rewards using the MaxEnt IRL algorithm.

Table 1. The rewards for each state (R(NC), R(U/B), R(IC/MA)) and action (R(LDCT/MG), R(I)) computed using the MaxEnt IRL algorithm, for one of the folds of the 5-fold cross validation, with an adaptive step size.

Normalization	R(NC)	R(U/B)	R(IC)	R(LDCT/M)	R(I)
Lung cancer					
None	83.530	127.410	−835.730	497.610	−427.530
By range	0.080	0.120	−0.800	0.540	−0.460
[0,1]	0.950	1.000	0.000	1.000	0.000
[−1,1]	0.910	1.000	−1.000	1.000	−1.000
Breast cancer					
None	−37.930	103.950	−571.420	−0.840	−1179.820
By range	−0.050	0.150	−0.800	−0.001	−0.999
[0,1]	0.790	1.000	0.000	1.000	0.000
[−1,1]	0.580	1.000	−1.000	1.000	−1.000

4.1 Comparison of MaxEnt IRL with and Without Adaptive Step Size

Table 1 shows the reward value of each state and action as well as different normalizations of these rewards computed using the MaxEnt IRL algorithm with an adaptive step size. We compare the MaxEnt IRL with and without the adaptive step size and assess the speed of convergence. Figure 3 depicts the computed rewards for states and actions for the lung POMDP over the number of iterations of gradient descent in the MaxEnt IRL algorithm, with and without an adaptive step size. A similar convergence trend is observed with the breast POMDP. As shown, the adaptive step size method converges to the correct solution more quickly than the standard MaxEnt IRL implementation. For the evaluation of the two models we use a reward function derived from rewards normalized in the $[-1,1]$ range.

4.2 Lung and Breast POMDP Results

We used the longitudinal observations from the NLST and Athena datasets as input to POMDPs such that each sequential observation updates the belief state of the agent. The belief state of the POMDP, at each epoch, is then used to select the next (optimal) action, with the objective of early detection of cancer. The POMDP models can suggest to continue screening (i.e., MG, LDCT) or to perform an intervention (i.e., biopsy or diagnostic imaging). If an intervention is performed, the individual is removed from further consideration. Evaluation of the POMDP is posed as a binary problem: if the POMDP suggests continued screening (LDCT/MG) then the patient is classified as a *negative* cancer; if it suggests an intervention, then the patient is classified as a *positive* cancer. Based on this definition, if the model suggests a LDCT/MG and the patient did not have a confirmed diagnosis of cancer in a given epoch, it is considered a true negative (TN); if the patient had a confirmed diagnosis of cancer then it is a false negative (FN). Conversely, if the model suggests an intervention and the patient did not have cancer in a given epoch, then it is considered a false positive (FP); if the patient had a diagnosis of cancer then it is considered a true positive (TP). Performance metrics were estimated for each epoch of the screening process. Any subject diagnosed with cancer is removed from the subsequent epoch. The POMDP models are compared against the equivalent physician decisions (recommendations) at each epoch, applying a similar framework for TN/FN/FP/TP to the experts, given the known cancer outcomes from each dataset (e.g., if the physicians suggested an LDCT/MG and the patient did not have a confirmed diagnosis of cancer, it is considered a true negative, etc.). Table 2 shows the performance of the lung and breast POMDPs and the corresponding performance of physicians on the same dataset. Notably, both POMDP models show performance comparable to experts. The lung cancer screening model has worse performance in terms of recall in the first and third screening epochs, but an improved performance in terms of recall and false positive rate in the second screening and post-screening. The breast cancer screening model demonstrates

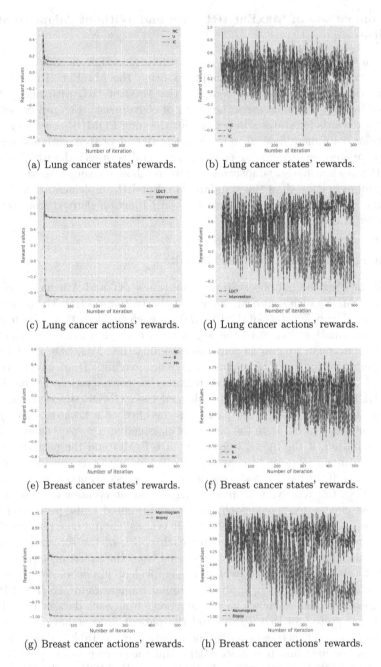

(a) Lung cancer states' rewards.   (b) Lung cancer states' rewards.

(c) Lung cancer actions' rewards.   (d) Lung cancer actions' rewards.

(e) Breast cancer states' rewards.   (f) Breast cancer states' rewards.

(g) Breast cancer actions' rewards.   (h) Breast cancer actions' rewards.

Fig. 3. State and action rewards computed using the MaxEnt IRL and normalized by range. **Left**: Using an adaptive step size. **Right**: Without using an adaptive step size. The adaptive step size MaxEnt IRL algorithm converges to a solution significantly faster than the MaxEnt IRL without an adaptive step size.

excellent recall (as do the expert physicians) but slightly worse false positive rate. The Cohen's kappa coefficient of agreement was used to assess the concordance between the POMDP models and physicians. The kappa score of the lung POMDP and physicians decreases over time due to the large number of false positives. A large portion of different cases are classified as false positives between the lung POMDP and physicians. The breast POMDP has a high kappa score demonstrating strong agreement with physicians in terms of false positives and true positives. For both lung and breast models, the variance of kappa per screening is less than 0.03.

Table 2. Left: The lung and breast POMDPs performance per epoch. **Right**: The physicians performance at each epoch. Metrics used for this evaluation are the true positive rate (TP), false negative rate (FN), false positive rate (FP) true negative rate (TN), precision (P), and recall (R). NCs: no-cancer cases. Cs: cancer cases. Kappa: Cohen's kappa score (coefficient of agreement), variance of kappa for all scores: < 0.03.

	POMDP						Physicians						Kappa
	TN rate	FP rate	FN rate	TP rate	Precision	Recall	TN rate	FP rate	FN rate	TP rate	Precision	Recall	
Lung cancer													
Training	NCs: 3960, Cs: Scr1, 2, 3 = 130, 68, 86; Pst-Scr = 78												
Scr 1	0.48	0.52	0.02	0.98	0.05	0.98	0.48	0.52	0.00	1.00	0.06	1.00	0.42
Scr 2	0.34	0.66	0.02	0.98	0.02	0.98	0.34	0.67	0.05	0.95	0.02	0.95	0.29
Scr 3	0.24	0.76	0.01	0.99	0.03	0.99	0.21	0.79	0.00	1.00	0.03	1.00	0.05
Pst-Scr	0.25	0.75	0.07	0.93	0.02	0.93	0.22	0.78	0.14	0.86	0.02	0.86	0.05
Testing	NCs: 990, Cs: Scr1, 2, 3 = 32, 17, 21; Pst-Scr = 20												
Scr 1	0.48	0.52	0.04	0.96	0.05	0.96	0.48	0.52	0.00	1.00	0.06	1.00	0.42
Scr 2	0.35	0.65	0.02	0.98	0.02	0.98	0.33	0.67	0.05	0.95	0.02	0.95	0.30
Scr 3	0.25	0.75	0.05	0.95	0.03	0.97	0.21	0.79	0.00	1.00	0.03	1.00	0.07
Pst-Scr	0.25	0.75	0.07	0.93	0.02	0.93	0.22	0.78	0.14	0.86	0.02	0.86	0.06
Breast cancer													
Training	NCs: 2808, Cs: Scr1, 2, 3, 4 = 370, 68, 27, 5												
Scr 1	0.99	0.01	0.01	0.99	0.96	0.99	0.99	0.01	0.01	0.99	0.95	0.99	1.00
Scr 2	0.99	0.01	0.01	0.99	0.70	0.99	0.99	0.01	0.01	0.99	0.73	0.99	0.97
Scr 3	0.98	0.02	0.03	0.97	0.40	0.97	0.99	0.01	0.03	0.97	0.43	0.97	0.95
Scr 4	0.98	0.02	0.00	1.00	0.09	1.00	0.98	0.02	0.00	1.00	0.10	1.00	0.92
Testing	NCs: 703, Cs: Scr1, 2, 3, 4 = 93, 17, 7, 1												
Scr 1	0.99	0.01	0.01	0.99	0.96	0.99	0.99	0.01	0.01	0.99	0.99	0.99	1.00
Scr 2	0.99	0.01	0.01	0.99	0.70	0.99	0.99	0.01	0.01	0.99	0.74	0.99	0.97
Scr 3	0.99	0.01	0.03	0.97	0.40	0.97	0.99	0.01	0.03	0.97	0.44	0.97	0.95
Scr 4	0.98	0.02	0.00	1.00	0.09	1.00	0.98	0.02	0.00	1.00	0.10	1.00	0.91

5   Discussion

POMDPs, through the use of beliefs and a hidden state space, can overcome some of the limitations seen in other sequential decision making models used in cancer screening. For instance, given the uncertainty in diagnosing lung and breast cancer from imaging studies, we modeled a hidden cancer state space in three parts [19]: no-cancer, benign/indeterminate, and malignant/invasive cancer. Modeling the cancer state space with an additional state rather than a binary state space allows the distinction of lower risk individuals (i.e., no abnormalities) – who constitute a large portion of screening cases and thus result in highly imbalanced datasets – over medium (i.e., benign growth) and high risk individuals (i.e., malignant abnormality).

Driven by the need to define the reward function in these screening POMDPs, we explored the use of the MaxEnt IRL algorithm towards generation of state-action reward pairs. As noted earlier, cost and utility estimation are frequently adopted as reward functions in healthcare models. [11] uses the National Statistical services' costs of procedures to define reward functions, while QALYs and a lifetime mortality risk model [16] are common alternative approaches. However, cost has certain limitations as it does not generalize to the whole population equally, and does not reflect the importance of quality outcomes. Additionally, QALY data are scarce, and arguably expensive to collect [16]. In contrast, a reward function learned using the MaxEnt IRL algorithm aims to maximize the objective of state-action trajectories. In this work, we used the MaxEnt IRL algorithm to generate reward functions for lung and breast cancer screening POMDP models using experts retrospective decisions. We improved the speed and accuracy of convergence of the gradient descent optimization of the MaxEnt IRL algorithm using an adaptive step size. Moreover, we introduced a multiplicative model for representing state-action pairs as products of state rewards and action rewards. The multiplicative model has the advantage to clearly demonstrate the difference in utility between rewards of different actions, which is what drives decision recommendation. Rewards are thus learned based on the state-visitation frequency of each trajectory. In this context, states with fewer visitations across each trajectory earn the lowest reward (e.g., invasive or malignant cancer state), which is why only cancer and non-cancer cases with a complete trajectory are used to learn rewards in our framework. Modeling the expert's decisions with the MaxEnt IRL algorithm resulted in reward functions for the POMDP models with performance comparable to experts. We noticed that when using aggressive reward functions (i.e., identifying all cancer cases), the true positive rate exceeded physicians' true positive rate but at the expense of a higher false positive rate, which in clinical practice can translate into higher costs and unnecessary psychological burden on the patient. Including more observational variables, derived from medical images, in the screening process can overcome this trade-off between true positive and false positive rate. The overall true positive rate and false positive rate using our learned reward functions in the POMDPs is comparable to experts. Nonetheless, in some cases the experts had false negative cases, which is also captured by our approach. When compared with other machine learning algorithms at the baseline of the lung and breast paradigms the POMDP models demonstrate improved performance.

The kappa coefficient of agreement between the POMDP models and physicians is constantly high for the breast POMDP model, illustrating the discriminatory capability of BI-RADS score as an imaging observation. In our lung cancer screening model, kappa gradually decreased over ensuing epochs, suggesting variability in the interpretation of LDCT imaging observations between the POMDP and the physicians. The lung POMDP is not fully replicating physicians' decision making patterns despite its overall performance being comparable to experts. When it comes to early cancer prediction (e.g., predicting screening 3 cancer from screening 1) the lung POMDP outperforms physicians, suggesting that the model and reward function are discriminating in a different way between positive and negative cases. Error analysis of the lung POMDP false positives shows a different subset from the physicians.

MaxEnt IRL also handles partial trajectories, making it suitable for screening processes in which individuals diagnosed with the disease exit the screening process for treatment. Relative to other IRL methods, MaxEnt IRL has the advantage of handling ambiguity by using a probabilistic model of behavior that exponentially prefers trajectories of higher reward [7,26]. MaxEnt IRL can also be used to transfer knowledge between datasets, tasks or domains by reusing learned weights (i.e., transfer learning). The only "partial" trajectory cases employed, in this analysis, are individuals diagnosed with cancer across the horizon of the screening process.

The first limitation of using MaxEnt IRL in this study is the fact that more than one combination of rewards can define the same problem. To overcome this, a policy iteration algorithm can be used rather than value iteration algorithm to learn optimal policies, as the policy space is finite in comparison to the rewards space (hence the policy iteration algorithm is guaranteed to optimally converge). A second limitation is the assumption that reward functions are only based on state visitation frequencies. The utility of screening recommendations is subjective and defined by different factors such as cost, quality of life, and patient satisfaction. To assess the quality of these reward functions a comparison of suggested recommendations with patient satisfaction could be used.

Other limitations are around assumptions about the nature of our datasets. While lung and breast cancer screening tests occurred roughly at one year intervals, we assumed that screening occurs annually (i.e., at fixed frequency). Moreover, data imbalance is a function of time, as at each screening point the number of cancer and non-cancer cases changes (i.e., at the outset of a screening period, more cancers are found at the beginning of a dataset). We did not account for this dynamic nature of the dataset during training. Given the small number of cancer cases across each screening point of both datasets, we utilized a stratified 5-fold cross-validation to obtain an unbiased estimate of model performance. Similarly, other temporal studies have used a k-fold cross validation to assess model performance [2,6,8,14,19,24]. To simplify modeling, our lung POMDP model considered only cases reporting a single pulmonary nodule over the course of the trial; this represents only a subset of the screened individuals, as many subjects have more than one such finding. A more concrete analysis would include cases with multiple nodules over time. However, it was not possible to ascertain the history of individual nodules in patients with multiple nodules as tracking of the nodules was not considered at the time of the study. Lastly, for the Athena dataset, in breast cancer screening, patients with BI-RADS 1, 2, or 3 rarely undergo biopsy, thus the true FN rate is likely underestimated. Future work involves the exploration of MaxEnt IRL in transfer learning between other datasets and domains, by reusing learned weights.

Acknowledgements. The authors thank the National Cancer Institute (NCI) for access to the National Lung Screening Trial data and Dr. Arash Naeim for access to the Athena Breast Health Network data collected at our institution. The authors would like to acknowledge the contribution of Dr. Audrey Winter for the preparation of the statistical evaluations, and Dr. William Speier for his comments on the paper. This material is based upon work supported by the National Science Foundation under Grant No. 1722516 and the Department of Radiological Sciences under the Data-Driven Diagnostic Decision Support (D4S) initiative.

References

1. Abbeel, P., Ng, A.Y.: Apprenticeship learning via inverse reinforcement learning. In: Twenty-First International Conference on Machine learning - ICML 2004, p. 1 (2004). https://doi.org/10.1145/1015330.1015430
2. Alaa, A.M., Moon, K.H., Hsu, W., Van Der Schaar, M.: ConfidentCare: a clinical decision support system for personalized breast cancer screening. IEEE Trans. Multimedia **18**(10), 1942–1955 (2016). https://doi.org/10.1109/TMM.2016.2589160, http://arxiv.org/abs/1602.00374
3. Alger, M.: Deep inverse reinforcement learning. Technical report (2016). https://matthewja.com/pdfs/irl.pdf
4. Babeş-Vroman, M., Marivate, V., Subramanian, K., Littman, M.: Apprenticeship learning about multiple intentions. In: Proceedings of the 28th International Conference on Machine Learning, ICML 2011, pp. 897–904 (2011)
5. Bennett, C.C., Hauser, K.: Artificial intelligence framework for simulating clinical decision-making: a Markov decision process approach. Artif. Intell. Med. **57**(1), 919 (2013). https://doi.org/10.1016/j.artmed.2012.12.003
6. Burnside, E.S., et al.: Probabilistic computer model developed from clinical data in national mammography database format to classify mammographic findings. Radiology **251**(3), 663–672 (2009). https://doi.org/10.1148/radiol.2513081346
7. Chelsea Finn: Deep RL Bootcamp Lecture 10B Inverse Reinforcement Learning - YouTube (2017). https://www.youtube.com/watch?v=d9DlQSJQAoI&t=1012s
8. Cuaya, G., et al.: A dynamic Bayesian network for estimating the risk of falls from real gait data. Med. Biol. Eng. Comput. **51**(1–2), 29–37 (2013). https://doi.org/10.1007/s11517-012-0960-2
9. D'Orsi, C.J.: ACR BI-RADS Atlas: Breast Imaging Reporting and Data System. American College of Radiology, Reston (2013)
10. Elson, S., Hiatt, R., Anton, C.: The Athena breast health network: developing a rapid learning system in breast cancer prevention, screening, treatment, and care. Breast Cancer Res. Treat. **140**, 417–425 (2013). https://doi.org/10.1007/s10549-013-2612-0

11. Goulionis, J.E., Vozikis, A., Benos, V.K., Nikolakis, D.: On the decision rules of cost-effective treatment for patients with diabetic foot syndrome. ClinicoEconomics Outcomes Res. **2**(1), 121–126 (2010). https://doi.org/10.2147/CEOR.S11981

12. Hauskrecht, M., Fraser, H.: Planning treatment of ischemic heart disease with partially observable Markov decision processes. Artif. Intell. Med. **18**(3), 221–244 (2000). https://doi.org/10.1016/S0933-3657(99)00042-1

13. Hauskrecht, M., Milos, H.: Dynamic decision making in stochastic partially observable medical domains: Ischemic heart disease example. In: Keravnou, E., Garbay, C., Baud, R., Wyatt, J. (eds.) AIME 1997. LNCS, pp. 296–299. Springer, Heidelberg (1997). https://doi.org/10.1007/bfb0029462

14. Van der Heijden, M., Velikova, M., Lucas, P.J.F.: Learning Bayesian networks for clinical time series analysis. J. Biomed. Inform. **48**, 94–105 (2014). https://doi.org/10.1016/j.jbi.2013.12.007

15. Klein, S., Pluim, J.P., Staring, M., Viergever, M.A.: Adaptive stochastic gradient descent optimisation for image registration. Int. J. Comput. Vis. **81**(3), 227–239 (2009). https://doi.org/10.1007/s11263-008-0168-y

16. Maillart, L.M., Ivy, J.S., Ransom, S., Diehl, K.: Assessing dynamic breast cancer screening policies. Oper. Res. **56**(6), 1411–1427 (2008). https://doi.org/10.1287/opre.1080.0614

17. National Lung Screening Trial Research Team, et al.: Reduced lung-cancer mortality with low-dose computed tomographic screening. N. Engl. J. Med. **365**(5), 395–409 (2011). https://doi.org/10.1056/NEJMoa1102873

18. Ng, A.Y., Russell, S.: Algorithms for inverse reinforcement learning. In: Proceedings of the Seventeenth International Conference on Machine Learning, pp. 663–670 (2000). https://doi.org/10.2460/ajvr.67.2.323

19. Petousis, P., Han, S.X., Aberle, D., Bui, A.A.: Prediction of lung cancer incidence on the low-dose computed tomography arm of the National Lung screening trial: a dynamic Bayesian network. Artif. Intell. Med. **72**, 42–55 (2016). https://doi.org/10.1016/j.artmed.2016.07.001

20. Schaefer, A.J., Bailey, M.D., Shechter, S.M., Roberts, M.S.: Modeling medical treatment using Markov decision processes. In: Brandeau, M.L., Sainfort, F., Pierskalla, W.P. (eds.) Operations Research and Health Care, pp. 597–616. Springer, Heidelberg (2005). https://doi.org/10.1007/1-4020-8066-2_23

21. Thrun, S., Burgard, W., Fox, D.: Probabilistic robotics (2006). https://doi.org/10.1145/504729.504754

22. Tusch, G.: Optimal sequential decisions in liver transplantation based on a POMDP model. In: ECAI, pp. 186–190 (2000)

23. Vroman, M.C.: Maximum likelihood inverse reinforcement learning. Ph.D. thesis (2014)

24. Watt, E.W., Bui, A.A.T.: Evaluation of a dynamic Bayesian belief network to predict osteoarthritic knee pain using data from the osteoarthritis initiative. In: AMIA 2008 Symposium, pp. 788–92 (2008). http://www.pubmedcentral.nih.gov/articlerender.fcgi?artid=2656041&tool=pmcentrez&rendertype=abstract

25. Ziebart, B.: Modeling purposeful adaptive behavior with the principle of maximum causal entropy. Ph.D. thesis (2010). http://www.cs.cmu.edu/~bziebart/publications/thesis-bziebart.pdf

26. Ziebart, B.D., Maas, A., Bagnell, J.A., Dey, A.K.: Maximum entropy inverse reinforcement learning. In: AAAI Conference on Artificial Intelligence, pp. 1433–1438 (2008)

Deep Learning Architectures for Vector Representations of Patients and Exploring Predictors of 30-Day Hospital Readmissions in Patients with Multiple Chronic Conditions

Muhammad Rafiq[1]([⊠]) , George Keel[1] , Pamela Mazzocato[1] ,
Jonas Spaak[1,2] , Carl Savage[1] , and Christian Guttmann[1,3,4]

[1] Department of Learning, Informatics, Management and Ethics (LIME),
Medical Management Centre, Karolinska Institutet, 171 65 Stockholm, Sweden
{muhammad.rafiq,george.keel,pamela.mazzocato,
jonas.spaak,carl.savage,christian.guttmann}@ki.se
[2] Department of Clinical Sciences, Danderyd University Hospital,
Karolinska Institutet, 182 88 Stockholm, Sweden
[3] Tieto Sweden AB, Fjärde Bassängvägen 15, 115 83 Stockholm, Sweden
[4] Nordic Artificial Intelligence Institute, Hälsingegatan 45,
113 31 Stockholm, Sweden

Abstract. This empirical study of a complex group of patients with multiple chronic concurrent conditions (diabetes, cardiovascular and kidney diseases) explores the use of deep learning architectures to identify patient segments and contributing factors to 30-day hospital readmissions. We implemented Convolutional Neural Network (CNN) and Recurrent Neural Network (RNN) on sequential Electronic Health Records data at the Danderyd Hospital in Stockholm, Sweden. Three distinct sub-types of patient groups were identified: chronic obstructive pulmonary disease, kidney transplant, and paroxysmal ventricular tachycardia. The CNN learned about vector representations of patients, but the RNN was better able to identify and quantify key contributors to readmission such as myocardial infarction and echocardiography. We suggest that vector representations of patients with deep learning should precede predictive modeling of complex patients. The approach also has potential implications for supporting care delivery, care design and clinical decision-making.

Keywords: 30-day hospital readmissions · Multiple Chronic Conditions ·
Deep learning

1 Introduction

Machine learning (ML) algorithms, particularly deep neural networks for sequential Electronic Health Records (EHR) data, have been extensively applied in the past decade to inform clinical decision making. However, the unstructured nature of EHR poses a challenge in its implementation, even more so if implemented for complex patients such as patients with Multiple Chronic Conditions (MCCs). The prevalence of

© Springer Nature Switzerland AG 2019
F. Koch et al. (Eds.): AIH 2018, LNAI 11326, pp. 228–244, 2019.
https://doi.org/10.1007/978-3-030-12738-1_17

patients with MCCs is increasing worldwide. In Sweden, patients with MCCs constitute 56.3% among adults between 35 to 75 years of age [1]. A prevalence study in Sweden found that 55% of the patient population had MCCs [2]. Similarly, in a population based longitudinal study in Stockholm, Sweden it was found that 33.6% of the participants had developed MCCs over a 3 years period [3]. In the US, the prevalence of patients with MCCs has increased from 21.8% in 2001 to 26.0% in 2010 [4]. The increasing number of concurrent chronic conditions are directly associated with the increase in health care costs [5]. In the US, the average per capita Medicare costs for patients with MCCs increased by 108.2% and 117.6% for concurrent 2 and 3 chronic conditions, respectively [6].

One such group of patients with MCCs is patients with concurrent diagnoses of cardiovascular and chronic kidney diseases and diabetes, hereinafter referred to as MCC patients. This triad of diseases constitutes a huge burden of disease around the world [7] due to high health care utilization [8]. MCC patients are complex due to the underlying pathophysiological mechanisms, conflicting treatment guidelines for each individual disease, and lack of studies [9, 10]. This makes it challenging for clinicians to treat MCC patients optimally.

Clinicians have been using deep neural networks such as Convolutional Neural Networks (CNN) and Recurrent Neural Networks (RNN) together with patients' medical histories and demographics (e.g. age and gender) to gain insights into the EHR data and tailor treatments according to individual patient needs. Sequential and time-dependency features in patients' journeys such as diagnoses and clinical procedures are increasingly being utilized in ML algorithm development to ensure accuracy and generalizability [11]. However, interpretation of the results obtained from deep neural networks is difficult. Practitioners tend to use simpler models as they have better interpretability, even though they are less accurate than modern ML algorithms [12]. This tradeoff between accuracy and interpretability of the model is not an optimal solution and researchers are developing modern ML algorithms that have both better accuracy and interpretability [13].

The aim of this study is to demonstrate how diagnosis and procedure codes contribute towards predicting 30-day hospital readmissions for MCC patients, and to explore MCC patients' sub-types through vector representations.

Our contribution through this study is three-fold: Firstly, based on the data obtained from one of the busiest tertiary hospitals in the Nordic countries, we demonstrate the effectiveness of deep learning architectures in the exploration of descriptive analytics for MCC patients. More specifically, we explore patterns in vector representations of patients and identify the contributors to 30-day hospital readmissions in terms of diagnoses and procedures. Secondly, the use of the Word2Vec model in conjunction with CNN places the MCC patients' records in EHR in sequential order for the entire care episode. This output can be fed into any type of deep neural architecture and used for making exploratory analysis and predictions. Thirdly, by applying the deep neural network architectures of CNN and RNN on real patient data set we demonstrate how these algorithms developed in one setting can be generalized and implemented in another setting.

2 Related Work

Hospital readmissions prediction, a key measure to assess quality of health care delivery [14–16], has increasingly become the focus of ML applications. This increasing focus is motivated by the highly incurred costs due to hospital readmissions [17] and their negative effect on the quality of patients' lives [18]. According to one estimate, the unplanned hospital readmissions cost $17.9 billion per year to the US health care system [19]. A recent estimate suggests that the costs related to hospital readmissions is around $26 billion per year for the US health care system [20].

Traditionally, demographics are used to predict hospital readmissions, but the use of other EHR variables is increasing with ML applications. A recent systematic review of models predicting hospital readmissions found that among the twenty-eight types of predictive risk variables, mostly comorbidity, demographics, and social variables were used [21]. Of the seventy-three unique predictive models, forty-five used socio-demographic variables, and only fourteen and sixteen used diagnoses and procedures respectively [21]. Among the twenty-two models related to cardiovascular diseases, only four used diagnoses together with the socio-demographic variables [22–25], seven used procedures with the socio-demographics variables [26–32], and none of them used diagnoses and procedures simultaneously. Two models used diagnoses and procedures simultaneously, but they were used to predict all-cause hospital readmissions [33, 34].

Most of the existing models for hospital readmissions prediction make use of the conventional ML algorithms such as logistic regression, support vector machines, and k-nearest neighbors [20, 21]. One potential limitation of the conventional approaches is that time duration and temporality are not taken into account which are an important aspect of building accurate prediction models. Ignoring temporality in prediction models based on EHR can lead to sub-optimal results [35].

RNNs have been recently implemented in health care to address issues of unequal time duration and temporality [11]. CNN and RNNs have been used to predict mortality [36], clinical events [35], diagnoses [37, 38], and clinical intervention [39]. CNN, a variant of RNN among others that are well known for capturing underlying structures in sequential data, have been applied in many fields such as speech recognition [40], natural language processing [41] and text classification [42]. Deep neural networks have also been recently implemented to predict hospital readmissions [20, 43]. However, only a few studies have implemented deep neural networks for exploratory and descriptive analysis of complex patients with MCCs [44].

3 Methods

3.1 Study Design

In this empirical study, we implemented two different types of RNNs to learn about vector representations of patients and factors contributing to 30-day hospital readmissions for MCC patients.

The literature has widely used 30-day hospital readmission as a performance measurement [14–16]. All the hospital admissions occurring within 30 days of the previous admission were considered as 30-day hospital readmissions regardless of the cause. Since readmissions shortly after hospital discharge add to clinical and financial burden [14, 45], we chose the measure as an initial metric in our exploration of predictive analytics for clinical planning and care coordination around MCC patients. Hospital readmissions can be either acute or non-acute depending on the patient's clinical condition and the treatment requirements. Even though acute readmissions are more expensive than non-acute, both types were considered for this study regardless of the underlying clinical reasons. Similarly, exploring MCC patients' sub-types through patient representations was chosen because it plays an important role in helping health care process analysis and operational improvement.

Experiment 1

In experiment 1, we implemented a CNN model complemented with the Word2Vec feature embedding in an unsupervised way, i.e. MCC patients readmitted within 30 days were not labelled and the algorithm learned the inherent structure from the EHR data. The study follows the approach developed by Zhao et al. [46].

We implemented the Word2Vec model that takes the patients' diagnoses (ICD codes) and procedure codes as input (in the form of a text corpus) and produced words in the form of output vectors. The Word2Vec model placed the ICD codes and procedures in their respective clusters. For MCC patients, the ICD codes and procedures were observed in a temporal order and they were organized into sequences. The Word2Vec model was trained on these sequences, and arrays of sequences were produced that were later fed into the CNN model in the form of a stacked matrix. The same hyper-parameters from Zhao et al. [46] were used.

The CNN model was trained on the learned sequences for MCC patients in the Word2Vec model. Inputs for MCC patients (p) were developed as an embedding matrix ($X_p \in R^{np \times d}$), where np is the number of records for MCC patients and d is the embedding dimension for MCC patients' diagnosis and procedure codes. A 1D convolution was applied over sequential dimension of the matrix. K filters were used in varying lengths of 2 to 5 to capture sequential variations. The filters looked for the presence of a specific pattern in the MCC patients' data. Max pooling layers were used that transformed user outputs from each filter to real numbers [46].

After the model was trained on MCC patients, a predominant single representation of MCC patients' sequential encounters was obtained which was further used to cluster MCC patients' vector representations. Various clustering algorithms were explored to group MCC patients' vector representations after learning the powerful sequential representation from the CNN model. Both feature based clustering (K-Means) and t-Distributed Stochastic Neighbor Embedding (tSNE) were used to cluster the MCC patients' vector representations.

Experiment 2

In experiment 2, we implemented the REverse Time AttentIoN (RETAIN) model, a variant of RNN, to identify the contributors of 30-day hospital readmissions for MCC patients. The study follows the approach developed by Choi et al. [35].

Using the RETAIN model we identified the predictors among diagnosis and procedure codes by assigning a significant portion of prediction to the attention weight generation process. EHR data is stored in such a way that every visit is recorded for each patient at a particular time and all the events that occur are recorded as multiple variables for each visit. The RETAIN model uses weights both for visits and for all the events occurring at a single visit, i.e. visit-level and variable-level weights. The visit level attention weights tackle the effect of each patient's visit embedding and variable level attention weights tackle each activity/event at a visit. For this purpose, RETAIN uses a model with two RNNs, i.e. RNNα and RNNβ, in a backwards direction to generate attention vectors. We used different hyper-parameters compared to Choi et al. [35], as shown in Table 1.

Table 1. Comparison of hyper-parameters used in RETAIN and Experiment 2.

Hyper-parameter	REATIN	Experiment 2
Size of visit embeddings, hidden layer for RNNα and RNNβ	128	4
Regularization for final classifier weight, input embedding weight, α generating weights, and β generating weights	0.0001	0.00001
Number of epochs	10	500
Training, validation and test split ratio	0.75:0.1:0.15	0.6:0.2:0.2

3.2   Data Collection

The study was conducted at an Integrated Multidisciplinary Clinic (HND-centrum) at a tertiary academic medical hospital, Danderyd University Hospital (DSAB), in Stockholm, Sweden. DSAB is one of the largest Emergency Hospitals (approximately 500 beds) in northern part of Stockholm, Sweden, and provides health services for approximately 650,000 people in eleven municipalities, mainly focused on internal medicine, cardiology, orthopedics, obstetrics and gynecology, and surgery and urology. The hospital is engaged in academic medical training and research, and is the first hospital in Sweden to open an Integrated Multidisciplinary Clinic to treat patients with MCCs of diabetes, cardiovascular and kidney diseases. Patients who were registered at the HND-centrum served as the participants of the study. We also included all patients that were referred to but not admitted to HND-centrum. These patients were included because they were the closest match of the patients registered at HND-centrum. These patients had to fulfil the inclusion criteria to be registered at the HND-centrum.

EHR data were obtained for all patients (n = 610) who were registered at the HND-centrum at DSAB between November 2010 and January 2017, and included personal identification numbers, visit dates, visit types, ICD diagnoses codes, clinical procedures codes, mortality dates, DRG codes, hospital admission dates and hospital discharge dates. All the data was anonymized before any preprocessing and patients were given a unique patient identification number. All the rows with missing values in the inpatient visits, outpatient visits and ICD codes columns were removed. Since the focus of this article was to develop HND patients' representations and identify predictors of 30-day hospital readmissions based only on the ICD codes (that would represent the disease progression for HND patients) and procedure codes, cleaned columns with patient identification numbers, ICD codes (diagnoses) and procedure codes were used in the subsequent analyses. The EHR data contained multiple entries of visits dates for the same patient on the same day, and therefore all the duplicate rows were removed. The index dates for 30-day hospital readmission were extracted from the data.

4 Results

The majority of MCC patients were between 70 and 80 years old. In total, 3,200 hospital admissions were observed in the selected period, the majority (87.5%) of which were acute hospital admissions (n = 2,801). A total of 76 diagnoses and 59 procedures were fed into the model to develop the sequence vector.

4.1 Experiment 1: CNN for Vector Representations of MCC Patients with 30-Day Hospital Readmissions

A total of 268 MCC patients with 30-day hospital readmissions were selected. The most significant ICD codes and procedures that contributed to the 30-day hospital readmissions were identified as shown in Table 2.

Table 2. Salient contributors (diagnoses and procedures) of 30-day hospital readmissions.

No	ICD code (diagnoses)	Procedures
1	I109 (essential hypertension)	Patient conference
2	E118 (diabetes mellitus type2)	Pacemaker control and reprogramming
3	N183 (chronic kidney disease stage 3)	Patient information and teaching
4	I259 (chronic ischemic heart disease)	Transthoracic Doppler echocardiography
5	I509 (heart failure)	Telemetry monitoring
6	N184 (chronic kidney disease stage 4)	Unplanned admission for end of life care
7	E119 (diabetes mellitus without complications)	Patient information and teaching
8	E117 (non-insulin-dependent diabetes)	Coronary angiography
9	Z921 (long term use of anticoagulants)	Use of interpreter
10	E785 (hyperlipidemia, unspecified)	Distant consultation

Table 2 shows that MCC patients readmitted to hospital within 30 days mostly had essential hypertension, diabetes mellitus type 2, and chronic kidney diseases stage 3 and 4 as the key contributors. Similarly, the most significant procedures that MCC patients experienced were patient conferences, echocardiography, coronary angiography, telemetry monitoring, and distant consultations. Since MCC patients are a niche group selected through very strict inclusion and exclusion criteria, it is not surprising that the model identified the contributors shown in Table 2, because these are the most frequent ICD codes and procedures among the selected MCC patients. However, the order of the contributions is noteworthy as they are ordered from most relevant to the least relevant.

Based on MCC patients' vector representations learned by the model, further exploration of the readmitted MCC patients was conducted. Distinct sub-types were identified by t-SNE clustering as shown in Fig. 1. Patients in cluster 1 appear to diverge most from the others.

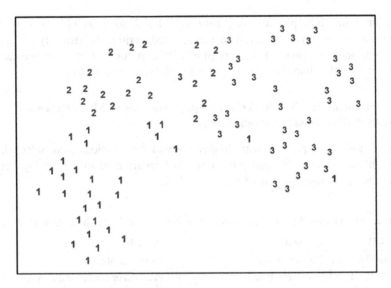

Fig. 1. t-SNE visualization of MCC patients' clusters. Each cluster represents sub-types within MCC patients who were readmitted within 30 days of their previous hospital admission.

In order to study what differentiates the three clusters from each other, we identified events most common to each cluster. Table 3 presents the key features for each cluster in terms of diagnoses and procedures that the MCC patients experienced (direct output from the model).

As shown in Table 3, Cluster 1 is distinct from Cluster 2 and 3. It was found that MCC patients in cluster 1 had Chronic Obstructive Pulmonary Disease (COPD), labelled with the ICD code R060 (Dyspnea), and required bronchodilation and spirometry more frequently. This indicates that MCC patients with concomitant COPD are more likely to be readmitted within 30 days.

MCC patients in Cluster 2 have had a kidney transplant (ICD code Z940) and went through several team visits and team conferences, illustrating the complex nature of their conditions.

Lastly, patients in Clusters 3 had paroxysmal ventricular tachycardia (ICD code I472) and repeatedly required control and reprogramming of their pacemaker or defibrillator, and hence were more likely to be readmitted within 30 days.

As we can see from the results in experiment 1, the model identified the most relevant contributing factors to 30-day hospital readmission among MCC patients, and the salient features of the clusters. However, the interpretation of the results can be very tricky because the contributing factors were not quantified or labelled as positive or negative. Thus, we aimed for a more explanatory model in experiment 2.

Table 3. Salient features of clusters of MCC patients with 30-day hospital readmissions.

Cluster 1	Cluster 2	Cluster 3
E117, N185, I350, I259, I109, E119, Cystoscopy, Information and teaching directed at patients, sampling (non-specific), R060, Arterial puncture, N409, Arterial puncture, Spirometry before and after bronchodilation, Exercise ECG standard, Spirometry before and after bronchodilation	N183, E107, Distant consultation, Z940, L979, N184, E117, I350, Patient conference, Team visit, Z940, Patient conference, E107, L979, Team visit, Z921, E117, L979, Patient conference	N183, E117, N184, Control and reprogramming of the pacemaker or defibrillator (AICD), E119, I109, N185, I509, Patient conference, Information and teaching directed at patients, N183, I472, Control and reprogramming of the pacemaker or defibrillator (AICD)

E117 = type 2 diabetes mellitus with multiple complications, N185 = chronic kidney disease stage 5, I350 = non-rheumatic aortic valve stenosis, I259 = nonspecific chronic ischemic heart disease, I109 = type 1 diabetes mellitus without complications, R060 = dyspnea, N409 = benign prostatic hyperplasia, N183 = chronic kidney disease stage 3, E107 = type 1 diabetes mellitus with multiple complications, Z940 = kidney transplant, L979 = non-pressure chronic ulcer of unspecified part of lower leg, N184 = chronic kidney disease stage 4, Z921 = long term use of blood thinning agents, E119 = type 2 diabetes mellitus without complications, I509 = unspecified heart failure, I472 = ventricular tachycardia

4.2 Experiment 2: RETAIN for MCC Patients with 30-Day Hospital Readmissions

Given the temporal sequences of MCC patients' diagnoses and procedures, we attempted to identify factors that contributed to 30-day hospital readmission. The same number of patients with 30-day hospital readmissions ($n = 268$) were included. We were able to attain Validation and Test accuracy of 0.900 and 0.794 respectively with RETAIN, which was better than the AUC of 0.8705 obtained by Choi et al. [35].

Table 4 shows results obtained from the RETAIN model in terms of contributions of the diagnoses and procedures to 30-day hospital readmissions and overall risk of readmission for MCC patients (direct output from the model). The contribution scores range between the lowest and highest values of -0.5 and 1.5 respectively. The overall risk score is calculated between 0 (no risk) and 1 (absolute risk). Table 4 shows readmission risk scores for three MCC patients and the contribution scores, either positive or negative, of each ICD code and procedure. The contribution scores show how each diagnosis and procedure contributed to the final prediction score (the contribution scores are added and put through the sigmoid function in the model).

Table 4. Contributions of the diagnosis and procedure codes in predicting 30-day hospital readmissions for MCC patients at each successive patient visit.

Visit No.	MCC patient 1	MCC patient 2	MCC patient 3
1	I489B: 0.353787	E107: −0.055551	E119: 0.651345
2	Z950: 1.018832	Allogenic red cell transfusion: 0.104865	Patient conference: 0.846334
3	I472: 0.638967	G473: 2.641692	Information and teaching directed at patients: 1.341413
4	Control and reprogramming of the pacemaker or defibrillator: −0.009479	Z921: 1.971326	Z921: 0.978889
5	Preoperative assessment: 0.073104	I489B: 0.557335	Patient conference: 0.196036
6	E785: 1.050597	I219: 0.019547	Distant consultation: 0.082917
7	I489B: 0.294173	I501: 0.006919	E119: 0.604113
8	G473: −0.223236	Telemetry monitoring: 0.004798	E669: 0.070579
9	Telemetry monitoring: −0.030263	Transthoracic Doppler echocardiography: −0.087382	N183: 0.365141
10	–	Coronary angiography: −0.042835	Orthostatic test: −0.021256
11	–	–	Distant consultation: 0.014503
12	–	–	I109: −0.029082
	Overall risk score: 0.952125	Overall risk score: 0.992927	Overall risk score: 0.992786

I489B = unspecified atrial fibrillation and atrial flutter, Z950 = presence of cardiac pacemaker, I472 = ventricular tachycardia, E785 = unspecified hyperlipidemia, G473 = sleep apnea, E107 = type 1 diabetes mellitus with multiple complications, Z921 = long term use of blood thinning agents, I219 = unspecified acute myocardial infarction, I501 = left ventricular failure, E119 = type 2 diabetes without complications, E669 = unspecified obesity, N183 = chronic kidney disease stage 3, I109 = essential hypertension

As we can see in Table 4, RETAIN determined the 30-day hospital readmission risk score for each individual MCC patient based on the diagnoses and procedures in their respective past medical encounters. In contrast to the results obtained from experiment 1, RETAIN assigned each diagnosis and procedure its specific contribution score and determined the overall risk of 30-day hospital readmission. Each contribution can either negatively or positively affect the overall risk of readmission.

MCC patient 1 in Table 4, likely belonging to Cluster 3 in experiment 1 because the patient had paroxysmal ventricular tachycardia (ICD I472), required reprogramming and controlling of the pacemaker or defibrillator, and preoperative assessment among other procedures. We can see that paroxysmal ventricular tachycardia diagnosis increased the risk of readmission (0.638967) while the procedure for control and programming of the pacemaker reduced the risk of readmission (−0.009479). Other notable contributions for MCC patient 1 were preoperative assessment and telemetry monitoring which increased (0.073104) and decreased (−0.030263) the risk for 30-day hospital readmission respectively.

MCC patient 2 had coagulation disorder (ICD code Z921) and acute myocardial infarction (ICD I219) in the past, and also required a blood transfusion among other things. We can see in Table 4 that a myocardial infarction and blood transfusion contributed positively to the readmission score (0.019547 and 0.104865 respectively), whereas echocardiography and coronary angiography reduced the risk of readmission (−0.087382 and −0.042835 respectively). Similarly, we can see the contributing factors and scores for MCC patient 3. For example, other than the three common diagnoses underlying the MCC condition, the patient had obesity (ICD code E669), which positively contributed to 30-day hospital readmission (0.070579).

5 Discussion

5.1 Effectiveness of CNN and RNN for MCC Patients

Both the CNN and RNN used in this study identified the most salient predictors for 30-day hospital readmissions among MCC patients. Experiment 1 demonstrated that various distinct sub-types exist among MCC patients readmitted within 30 days, and MCC patients with COPD, kidney transplant and paroxysmal ventricular tachycardia are at higher risk of readmission within 30 days. Experiment 2 demonstrated that the model was able to identify contribution scores for diagnoses and procedures such as myocardial infarction and echocardiography, and overall 30-day hospital readmission risks for individual MCC patients.

Experiment 1 also demonstrated that patient conferences preceded 30-day hospital readmissions among MCC patients. Control and reprogramming of pacemakers, telemetry monitoring and distant consultations were also associated with the 30-day hospital readmissions. But, experiment 1 was not able to demonstrate the quantitative contribution, either positive or negative, of the diagnoses and procedures to 30-day hospital readmissions. However, this challenge associated with interpretation of results obtained from deep neural networks is common, and some models have been applied in health care to address this issue [35, 47].

In experiment 2, we demonstrated that RETAIN was better able to quantify the individual contributions of diagnoses and procedures to 30-day hospital readmissions, both in terms of either causing (positive contribution) or preventing (negative contribution) readmissions. These contributions were determined for individual patients considering the sequence and timing of the previous MCC patients' visits as shown in Table 4. In coming studies we aim to explore how the RETAIN model is affected by adding more variables to the MCC patients' sequential records, such as demographics, medications, laboratory values, number of visits, and length-of-stay.

5.2   Sub-typing and Vector Representations of MCC Patients

This study suggests that vector representations of patients and sub-typing among complex patients, such as MCC patients, by the type of health care encounters, like hospital readmissions, is as important as sub-typing patients by age and gender. This nontraditional approach to sub-typing makes it more robust and useful in practice. Sub-typing complex patients by the type of health care encounters enables us to have a deeper look at the sub-types as it adds up the possibility to include temporal aspects of the patients' encounters. The recently developed robust deep learning architectures have made it easier to include more variables from the EHR in predictive analytics and at the same time produce robust results.

Vector representations of patients are a road map in disease progression [48–50] that can identify key disease patterns among complex chronic patients. Patients with MCCs are difficult to treat because there are no clear guidelines for such patients. Majority of these guidelines are developed mostly based on the individual diseases. Vector representations of patients through deep learning architectures can be useful in identifying the unique patients that possess uniform and distinct features. These unique patients require unique approaches that need to be followed in their treatment processes.

Sub-typing patients, and hence identifying specific medical journeys for complex patients may enhance clinicians' ability to make optimal care decisions. Researchers are identifying sub-types among patients with single medical conditions, such as diabetes. These sub-types have unique features and require that they are treated differently. For instance, a recent study performing cluster analysis of diabetic patients found five different clusters with unique characteristics and risk profiles [51]. Patients with MCCs are stronger candidates for sub-typing since they have a combination of complex medical conditions and the underlying unique characteristics. Among patients with MCCs, the utilization of time aware and temporal variables may provide an opportunity to explore sub-types of patients with uniform and distinct features.

As demonstrated in this study, robust deep learning algorithms such as CNN and RNN have been proposed to learn typical vector representations of complex patients, and stratify them into suitable sub-types that can help clinicians in their day-to-day decision making process during disease and operational management. Such algorithms may improve the risk assessment for 30-day hospital readmissions and help clinicians and health care managers in planning their daily operations in an optimal manner.

5.3   Practical Implications for MCC Patients and Health Care in General

The CNN and RNN used in this study have the potential to positively influence practical decisions around MCC patients and optimize resource utilization. The models can be used to inform clinicians about high consumers of care, and develop process maps and clinical pathways for unique clusters among MCC patients and complex chronic patients in general.

Patients with MCCs have implications for health care costs [5]. Identification of high consumers of care is a priority issue for decision makers and plays a key role in the treatment and hospital resource planning. Among patients with MCCs, health care costs are differently associated with individual patients since they have different underlying conditions, as we have shown in this study. Therefore, it is imperative that some of the sub-types will incur more health care costs than the others. Deep learning architectures may help clinicians and health care managers to identify such costly patients with MCCs. Clinicians and health care managers can then develop individually tailored strategies and may help in reducing health care costs. For instance, by identifying patients with greater risks of being readmitted within 30 days, clinicians can develop preventative strategies to reduce the risk of readmissions. Similarly, the hospital can plan well in advance by better allocation of health care resources. The models can also be used in the development of individually tailored preventative and cost reducing strategies for different groups of patients with complex chronic conditions other than MCC patients.

This study has practical implications for some of the areas of knowledge representation in health care such as organization of knowledge for the management of MCC patients, computer based knowledge representation, development of knowledge based systems, diagnostic problem solving methodologies, and treatment and hospital resources planning.

Our study has implications for the organization of knowledge for the management of MCC patients. Deep learning architectures can prove to be useful in the development of process maps and clinical pathways. Such process maps and clinical pathways can be useful in the identification of sequential events happening in patients with a specific group of MCCs and also for the underlying specific sub-types. Since these architectures can learn the predominant sequential representations of health encounters throughout patients medical history, these underlying learned representations can be visualized into process maps and clinical pathways. These individually tailored process maps may prove to be useful in the development of clinical practice guidelines for patients with MCCs.

Our study also has implications for predictive analytics methodologies in clinical practice in general. The models implemented in this study have predictive potential and can be used to identify MCC patients' sub-types for which prediction models can be developed. These models can predict both clinical and healthcare operations management outcomes such as mortality, cardiovascular events, length-of-stay, and hospital readmissions. For instance, if clinicians are able to classify patients into a particular cluster that follows a specific sequence of treatment events, they would be able to stream patients into a sequential care process, or if not, into a more customized process [52]. In our future studies, we intend to improve the current models and potentially develop a predictive decision support model for clinicians at HND-centrum.

Lastly, our study serves as a preliminary step in the development of knowledge based systems and heterogeneous software applications integration in health care. The predictive models developed on the basis of the deep learning architectures in this study can be incorporated into interactive analytics tools for patients with MCCs where clinicians could review individual MCC patient's risk scores for selected outcomes. The tool can also provide a visualized overview of a patient's past medical history and encounters. Once we have improved and developed a predictive decision support model, we aim to develop such interactive analytics tool that can help clinicians and health care managers alike in their day to day activities. The tool will support clinicians in visualizing MCC patients' past medical history and their predominant clinical processes, and also predict various clinical outcomes. The tool will also help health care managers in health care operations management activities.

6 Limitations/Methodological Considerations

This study reports the findings based on ICD codes and procedures only for a relatively small sample size of MCC patients. Therefore, prediction score accuracy might be low. We aim to increase the sample size and refine the models' parameters in order to increase accuracy. Factors that might be important for predicting readmissions e.g. laboratory values were not considered due to the design of this study. We intend to include more variables for the development of a predictive decision support model for MCC patients. Some ICD codes and procedures are also inherently associated with certain already-made clinical decisions. In the future, we will carefully select diagnosis and procedure codes relevant for specified research questions, for instance using directed acyclic graphs. Lastly, since this study was conducted on MCC patients' data obtained from a single hospital, the generalizability of the findings to other settings may be limited.

7 Conclusion

Temporal data on ICD codes and procedures appears to be valuable for personalized disease management strategies for MCC patients. In this study, three distinct sub-types of MCC patients with increased risk for 30-day hospital readmission were identified i.e. MCC patients with chronic obstructive pulmonary disease, kidney transplant, and paroxysmal ventricular tachycardia. We suggest that temporal vector representations of patients and sub-typing with deep neural networks such as CNN and RNN are useful in the development of predictive analytics tools for patients with MCCs.

In the future, we plan to explore the application of deep neural networks in the development of prediction models for MCC patients that can be used to predict both clinical and healthcare operations management outcomes. We also aim to make the results easily accessible for clinicians and other health care professionals by developing interactive analytics tools.

Funding. This work was financially supported by Vårdalstiftelsen, with additional financial support from FORTE, the Kamprad Family Foundation, and Strategic Research Area Health Care Science. The author PM was funded by Karolinska Institutet/Umeå University during the project period. The funders had no involvement in the study design; in the collection, analysis and interpretation of the data; in the writing of the report; and in the decision to submit the paper for publication.

Ethical Considerations. The study has been approved by the Regional Ethics Committee (Diary Numbers: 2014/384-31/1 and 2017/999-31/2).

References

1. Pache, B., Vollenweider, P., Waeber, G., Marques-Vidal, P.: Prevalence of measured and reported multimorbidity in a representative sample of the Swiss population. BMC Public Health. **15**(1), 164 (2015). https://doi.org/10.1186/s12889-015-1515-x
2. Marengoni, A., Winblad, B., Karp, A., Fratiglioni, L.: Prevalence of chronic diseases and multimorbidity among the elderly population in Sweden. Am. J. Publ. Health **98**(7), 1198–1200 (2008). https://doi.org/10.2105/AJPH.2007.121137
3. Melis, R., Marengoni, A., Angleman, S., Fratiglioni, L.: Incidence and predictors of multimorbidity in the elderly: a population-based longitudinal study. PLoS ONE. **9**(7), e103120 (2014). https://doi.org/10.1371/journal.pone.0103120. (Ed. by, A. Scuteri)
4. Ward, B.W., Schiller, J.S.: Prevalence of multiple chronic conditions among US adults: estimates from the National Health Interview Survey, 2010. Prev. Chronic Dis. **25**(10), 120203 (2013). https://doi.org/10.5888/pcd10.120203
5. Sambamoorthi, U., Tan, X., Deb, A.: Multiple chronic conditions and healthcare costs among adults. Expert Rev. Pharmacoecon. Outcomes Res. **15**(5), 823–832 (2015). https://doi.org/10.1586/14737167.2015.1091730
6. Schneider, K.M., O'Donnell, B.E., Dean, D.: Prevalence of multiple chronic conditions in the United States' Medicare population. Health Qual. Life Outcomes **7**(1), 82 (2009). https://doi.org/10.1186/1477-7525-7-82
7. Suckling, R., Gallagher, H.: Chronic kidney disease, diabetes mellitus and cardiovascular disease: Risks and commonalities. J. Ren. Care **38**, 4–11 (2012). https://doi.org/10.1111/j.1755-6686.2012.00274.x
8. Johnson, T.L., et al.: For many patients who use large amounts of health care services, the need is intense yet temporary. Health Aff. (Millwood) **34**(8), 1312–1319 (2015). https://doi.org/10.1377/hlthaff.2014.1186
9. Marengoni, A., et al.: Aging with multimorbidity: a systematic review of the literature. Ageing Res. Rev. **10**(4), 430–439 (2011). https://doi.org/10.1016/j.arr.2011.03.003
10. Fortin, M., Lapointe, L., Hudon, C., Vanasse, A.: Multimorbidity is common to family practice: is it commonly researched? Can. Fam. Physician **51**, 244–245 (2005). PMID 16926936
11. Baytas, I.M., Xiao, C., Zhang, X., Wang, F., Jain, A.K., Zhou, J.: Patient subtyping via time-aware LSTM networks. In: Proceedings of 23rd ACM SIGKDD International Conference on Knowledge Discovery and Data Mining - KDD 2017, pp. 65–74 (2017). https://doi.org/10.1145/3097983.3097997
12. Caruana, R., Lou, Y., Gehrke, J., Koch, P., Sturm, M., Elhadad, N.: Intelligible models for healthcare. In: Proceedings of 21st ACM SIGKDD International Conference on Knowledge Discovery and Data Mining - KDD 2015, pp. 1721–30 (2015). https://doi.org/10.1145/2783258.2788613
13. Guttmann, C., Sun, X.Z.: Balancing provenance and accuracy tradeoffs in data modeling 2016. United States Patent No: US 9.275.425 B2
14. Bosco, J.A., Karkenny, A.J., Hutzler, L.H., Slover, J.D., Iorio, R.: Cost burden of 30-day readmissions following medicare total hip and knee arthroplasty. J. Arthroplasty **29**(5), 903–905 (2014). https://doi.org/10.1016/j.arth.2013.11.006

15. Stefan, M.S., et al.: Hospital performance measures and 30-day readmission rates. J. Gen. Intern. Med. **28**(3), 377–385 (2013). https://doi.org/10.1007/s11606-012-2229-8

16. Wish, J.B.: The role of 30-day readmission as a measure of quality. Clin. J. Am. Soc. Nephrol. **9**(3), 440–442 (2014). https://doi.org/10.2215/CJN.00240114

17. Basu Roy, S., et al.: Dynamic hierarchical classification for patient risk-of-readmission. In: Proceedings of the 21st ACM SIGKDD International Conference on Knowledge Discovery and Data Mining - KDD 2015, pp. 1691–700. ACM Press, New York (2015). https://doi.org/10.1145/2783258.2788585

18. McIlvennan, C.K., Eapen, Z.J., Allen, L.A.: Hospital readmissions reduction program. Circulation **131**(20), 1796–1803 (2015). https://doi.org/10.1161/CIRCULATIONAHA.114.010270

19. Jencks, S.F., Williams, M.V., Coleman, E.A.: Rehospitalizations among patients in the medicare fee-for-service program. N. Engl. J. Med. **360**(14), 1418–1428 (2009). https://doi.org/10.1056/NEJMsa0803563

20. Xie, J., Zhang, B., Ma, J., Zeng, D.D., Ciganic, J.L.: Readmission prediction for patients with heterogeneous hazard: a trajectory-based deep learning approach. SSRN Electron J. 1–41 (2018). https://doi.org/10.2139/ssrn.3144798

21. Zhou, H., Della, P.R., Roberts, P., Goh, L., Dhaliwal, S.S.: Utility of models to predict 28-day or 30-day unplanned hospital readmissions: an updated systematic review. BMJ Open **6**(6), e011060 (2016). https://doi.org/10.1136/bmjopen-2016-011060

22. Keyhani, S., Myers, L.J., Cheng, E., Hebert, P., Williams, L.S., Bravata, D.M.: Effect of clinical and social risk factors on hospital profiling for stroke readmission. Ann. Intern. Med. **161**(11), 775 (2014). https://doi.org/10.7326/M14-0361

23. Rana, S., Tran, T., Luo, W., Phung, D., Kennedy, R.L., Venkatesh, S.: Predicting unplanned readmission after myocardial infarction from routinely collected administrative hospital data. Aust. Health Rev. **38**(4), 377 (2014). https://doi.org/10.1071/AH14059

24. Donzé, J., Lipsitz, S., Schnipper, J.L.: Risk factors for potentially avoidable readmissions due to end-of-life care issues. J. Hosp. Med. **9**(5), 310–314 (2014). https://doi.org/10.1002/jhm.2173

25. Taha, M., Pal, A., Mahnken, J.D., Rigler, S.K.: Derivation and validation of a formula to estimate risk for 30-day readmission in medical patients. Int. J. Qual. Health Care **26**(3), 271–277 (2014). https://doi.org/10.1093/intqhc/mzu038

26. Hebert, C., et al.: Diagnosis-specific readmission risk prediction using electronic health data: a retrospective cohort study. BMC Med. Inform. Decis. Mak. **14**(1), 65 (2014). https://doi.org/10.1186/1472-6947-14-65

27. Iannuzzi, J.C., Chandra, A., Kelly, K.N., Rickles, A.S., Monson, J.R.T., Fleming, F.J.: Risk score for unplanned vascular readmissions. J. Vasc. Surg. **59**(5), 1340–1347.e1 (2014). https://doi.org/10.1016/j.jvs.2013.11.089

28. Lucas, D.J., et al.: Assessing readmission after general, vascular, and thoracic surgery using ACS-NSQIP. Ann. Surg. **258**(3), 430–439 (2013). https://doi.org/10.1097/SLA.0b013e3182a18fcc

29. Wallmann, R., Llorca, J., Gómez-Acebo, I., Ortega, Á.C., Roldan, F.R., Dierssen-Sotos, T.: Prediction of 30-day cardiac-related-emergency-readmissions using simple administrative hospital data. Int. J. Cardiol. **164**(2), 193–200 (2013). https://doi.org/10.1016/j.ijcard.2011.06.119

30. Wasfy, J.H., et al.: A prediction model to identify patients at high risk for 30-day readmission after percutaneous coronary intervention. Circ. Cardiovasc. Qual. Outcomes **6**(4), 429–435 (2013). https://doi.org/10.1161/CIRCOUTCOMES.111.000093

31. Raposeiras-Roubín, S., et al.: Mortality and cardiovascular morbidity within 30 days of discharge following acute coronary syndrome in a contemporary European cohort of patients: How can early risk prediction be improved? The six-month GRACE risk score. Rev. Port. Cardiol. **34**(6), 383–391 (2015). https://doi.org/10.1016/j.repc.2014.11.020

32. Sudhakar, S., Zhang, W., Kuo, Y.-F., Alghrouz, M., Barbajelata, A., Sharma, G.: Validation of the readmission risk score in heart failure patients at a tertiary hospital. J. Card. Fail. **21**(11), 885–891 (2015). https://doi.org/10.1016/j.cardfail.2015.07.010

33. van Walraven, C., Wong, J., Forster, A.J., Hawken, S.: Predicting post-discharge death or readmission: deterioration of model performance in population having multiple admissions per patient. J. Eval. Clin. Pract. **19**(6), 1012–1018 (2013). https://doi.org/10.1111/jep.12012

34. van Walraven, C., Wong, J., Forster, A.J.: LACE+index: extension of a validated index to predict early death or urgent readmission after hospital discharge using administrative data. Open Med. **6**(3), e80–e90 (2012). PMID 23696773

35. Choi, E., Bahadori, M.T., Kulas, J.A., et al.: RETAIN: an interpretable predictive model for healthcare using reverse time attention mechanism. In: NIPS (2016). arXiv:1608.05745

36. Aczon, M., et al.: Dynamic mortality risk predictions in pediatric critical care using recurrent neural networks, 1–18 (2017). arXiv:1701.06675

37. Lipton, Z.C., Kale, D.C., Elkan, C., Wetzel, R.: Learning to diagnose with LSTM recurrent neural networks, 1–18 (2015). https://doi.org/10.14722/ndss.2015.23268

38. Razavian, N., Marcus, J., Sontag, D.: Multi-task prediction of disease onsets from longitudinal lab test, 1–27 (2016). arXiv:1608.00647

39. Suresh, H., Hunt, N., Johnson, A., Celi, L.A., Szolovits, P., Ghassemi, M.: Clinical intervention prediction and understanding using deep networks, 1–16 (2017). arXiv:1705.08498

40. Graves, A., Mohamed, A., Hinton, G.: Speech recognition with deep recurrent neural networks, (3) (2013). https://doi.org/10.1109/icassp.2013.6638947

41. Wen, T.-H., Gasic, M., Mrksic, N., Su, P.-H., Vandyke, D., Young, S.: Semantically conditioned LSTM-based natural language generation for spoken dialogue systems (2015). arXiv:1508.01745

42. Lai, S., Xu, L., Liu, K., Zhao, J.: Recurrent convolutional neural networks for text classification. In: Twenty-Ninth AAAI Conference on Artificial Intelligence, pp. 2267–2273 (2015). ISBN 9781577357018

43. Xiao, C., Ma, T., Dieng, A.B., Blei, D.M., Wang, F.: Readmission prediction via deep contextual embedding of clinical concepts. PLoS ONE. **13**(4), e0195024 (2018). (Ed. by, C. Hou). https://doi.org/10.1371/journal.pone.0195024

44. Futoma, J., Sendak, M., Cameron, C.B., Heller, K.: Predicting disease progression with a model for multivariate longitudinal clinical data. J. Mach. Learn. Res. **56**, 42–54 (2016)

45. Joynt, K.E., Jha, A.K.: Thirty-day readmissions—truth and consequences. N. Engl. J. Med. **366**(15), 1366–1369 (2012). https://doi.org/10.1056/NEJMp1201598

46. Zhao, C., Shen, Y.: Convolutional neural network-based model for patient representation learning to uncover temporal phenotypes for heart failure (2017)

47. Che, Z., Purushotham, S., Khemani, R., Liu, Y.: Distilling knowledge from deep networks with applications to healthcare domain, 1–13 (2015). arXiv:1512.03542

48. Parr, D.G.: Patient phenotyping and early disease detection in chronic obstructive pulmonary disease. Proc. Am. Thorac. Soc. **8**(4), 338–349 (2011). https://doi.org/10.1513/pats.201101-014RM

49. Shickel, B., Tighe, P.J., Bihorac, A., Rashidi, P.: Deep EHR: a survey of recent advances in deep learning techniques for Electronic Health Record (EHR) analysis. IEEE J. Biomed. Health Inform. **22**(5), 1589–1604 (2018). https://doi.org/10.1109/JBHI.2017.2767063

50. Miotto, R., Li, L., Kidd, B.A., Dudley, J.T.: Deep patient: an unsupervised representation to predict the future of patients from the electronic health records. Sci. Rep. **6**(1), 26094 (2016). https://doi.org/10.1038/srep26094
51. Chatterjee, S., Davies, M.J.: Accurate diagnosis of diabetes mellitus and new paradigms of classification. Nat. Rev. Endocrinol. **14**(7), 383–384 (2018). https://doi.org/10.1038/s41574-018-0025-1
52. Bohmer, R.M.J.: Designing Care: Aligning the Nature and Management of Health Care. Harvard Business School Press, Brighton (2009). ISBN 142217560X

Author Index

Printed in the United States
By Bookmasters